Engineering Writing by Design

CREATING FORMAL DOCUMENTS OF LASTING VALUE

Edward J. Rothwell
Department of Electrical and Computer Engineering
Michigan State University, East Lansing, USA

Michael J. Cloud
Department of Electrical and Computer Engineering
Lawrence Technological University
Southfield, Michigan, USA

CRC Press
Taylor & Francis Group
Boca Raton   London   New York

CRC Press is an imprint of the
Taylor & Francis Group, an **informa** business

CRC Press
Taylor & Francis Group
6000 Broken Sound Parkway NW, Suite 300
Boca Raton, FL 33487-2742

© 2014 by Taylor & Francis Group, LLC
CRC Press is an imprint of Taylor & Francis Group, an Informa business

No claim to original U.S. Government works

Printed on acid-free paper
Version Date: 20140415

International Standard Book Number-13: 978-1-4822-3431-2 (Paperback)

This book contains information obtained from authentic and highly regarded sources. Reasonable efforts have been made to publish reliable data and information, but the author and publisher cannot assume responsibility for the validity of all materials or the consequences of their use. The authors and publishers have attempted to trace the copyright holders of all material reproduced in this publication and apologize to copyright holders if permission to publish in this form has not been obtained. If any copyright material has not been acknowledged please write and let us know so we may rectify in any future reprint.

Except as permitted under U.S. Copyright Law, no part of this book may be reprinted, reproduced, transmitted, or utilized in any form by any electronic, mechanical, or other means, now known or hereafter invented, including photocopying, microfilming, and recording, or in any information storage or retrieval system, without written permission from the publishers.

For permission to photocopy or use material electronically from this work, please access www.copyright.com (http://www.copyright.com/) or contact the Copyright Clearance Center, Inc. (CCC), 222 Rosewood Drive, Danvers, MA 01923, 978-750-8400. CCC is a not-for-profit organization that provides licenses and registration for a variety of users. For organizations that have been granted a photocopy license by the CCC, a separate system of payment has been arranged.

Trademark Notice: Product or corporate names may be trademarks or registered trademarks, and are used only for identification and explanation without intent to infringe.

**Visit the Taylor & Francis Web site at
http://www.taylorandfrancis.com**

**and the CRC Press Web site at
http://www.crcpress.com**

To Rick Nolan; my appreciation for your encouragement and support over a lifetime of friendship far exceeds what I can express in these few words — EJR

To Joan Fellows, a creative soul who listened with patience, insight, and acceptance — MJC

Contents

Preface

In *Walden*, Henry Thoreau asserts that "Books must be read as deliberately and reservedly as they were written." Given the serious and highly technical nature of formal engineering writing, any reader of such writing would be wise to follow Thoreau's advice. The purpose of the present book, however, is to speak to the engineering writer. Our basic premise is that engineering material should be written as deliberately and carefully as it will be read.

Engineers are smart people and their work is important. Their writing should not be inaccurate, vague, ambiguous, or otherwise opaque. To a great extent, modern engineering is an extension of science and mathematics and is therefore amenable to clear and logical exposition. Some aspects of engineering remain more art than science, to be sure. We would argue, however, that in such cases it is *especially* important for engineers to write precisely, as readers will be in less of a position to close expository gaps through deductive reasoning. In other words, *clear description* is just as important in technical writing as *clear argumentation*. Technical subjects can make for difficult reading as it is. A reader should not have to go over a passage again and again, finally being forced to guess whether the writer was attempting to motivate a viewpoint, describe something that already exists, describe something that could conceivably exist, draw a conclusion from known facts, persuade, or something else. Yet, a writer who approaches the writing task carelessly, by simply accumulating a pile of words and equations, may produce just that sort of confusion.

Our combined experience as engineering educators has determined the composition of this book. We routinely interact with students who not only lack writing ability but see it as unimportant or, at least, not worth the effort to attain. They miss the fact that, as engineers, they already have the tools to become good writers. By applying the same mindset to writing as they do to engineering design, they will discover that to write like engineers, they must think like engineers. Unfortunately, this key observation does not seem to be intuitively obvious. It has become increasingly apparent to us that such a viewpoint must be explicitly taught.

We believe that formal engineering writing can be taught (and learned) in the context of the modern engineering design process. A writing task can be seen and approached as a design problem accompanied by requirements, constraints, protocols and standards to meet, and an eventual customer to satisfy (the target reader). Engineers are deeply familiar with design-oriented thinking, and their experience with design processes can be brought directly to bear on even the largest writing tasks. It provides an alternative framework for driving home the classical elements of English composition: unity, coherence, and emphasis.

The general topic of "engineering writing" is obviously very broad. In the usage we adopt here, "formal" engineering writing would be aimed at the production of such documents as undergraduate term papers, capstone design reports, literature reviews, master's theses, doctoral dissertations, corporate technical reports, journal articles, books, and research or business proposals. Our purpose is not to address the composition of contracts, business letters, office memos, emails, faxes, Powerpoint slides, conference posters, meeting minutes, or pages in a laboratory notebook. These things are important — and certainly some of the principles we put forth will apply to the writing in *any* serious engineering document — but they are not our focus.

Acknowledgments

We would like to thank Beth Lannon-Cloud for her constructive feedback on the manuscript. It was Beth who pointed out the value of "negative" examples, explicitly showing the reader common mistakes to avoid. She suggested we include more of these, and we did. Thanks to Ramon E. Moore, founder of the branch of mathematics known as interval analysis, for commenting on an early draft of Chapter 7. Adena Moore suggested we break up some of our own long paragraphs (an example of the value of getting feedback on one's writing); she also provided counsel regarding gender neutrality in English. Leonid P. Lebedev, a widely published mathematician and mechanicist, offered valuable remarks that were implemented throughout the manuscript. We are indebted to Ben Crowgey and Junyan Tang for carefully reading the manuscript and providing many helpful comments. Thanks to Dan Pokora, Melissa Turner, Mary S. Lannon, Gordon P. Fellows, Mary Ann Lannon, Bernard Kulwicki, William Kolasa, Byron Drachman, Dennis Nyquist, and Lisa Anneberg for their kind encouragement. Steve Leuty, Maureen Lannon, Mary Jo Fellows, Gordie Lee Fellows, Eric Schreiber, and Nicole McDermott all assisted with the dedication page. We are grateful for the work of several anonymous reviewers. Our Taylor & Francis editors Nora Konopka, Michele Smith, Kate Gallo, and Michele Dimont provided much valuable guidance and support throughout the publication process. The copyeditor was Alice Mulhern and the cover designer was John Gandour.

Authors

Edward J. Rothwell earned a BS from Michigan Technological University, MS and EE from Stanford University, and PhD from Michigan State University, all in electrical engineering. He has been a faculty member in the Department of Electrical and Computer Engineering at Michigan State University since 1985, and currently holds the rank of professor. Before coming to Michigan State he worked at Raytheon and Lincoln Laboratory. Dr Rothwell has published numerous articles in professional journals involving his research in electromagnetics and related subjects. He is coauthor with Michael Cloud of *Electromagnetics* (CRC Press, 2nd ed., 2008). Dr Rothwell is a member of Phi Kappa Phi, Sigma Xi, URSI Commission B, and is a Fellow of the IEEE.

Michael J. Cloud was awarded a BS, MS, and PhD from Michigan State University, all in electrical engineering. He has been a faculty member in the Department of Electrical and Computer Engineering at Lawrence Technological University since 1987, and currently holds the rank of associate professor. Dr Cloud has coauthored eleven other books, mostly in engineering mathematics. He is a senior member of the IEEE.

To the Reader

This is a book on technical writing for engineers. There are numerous books on technical writing — some new, some very old. Many take hundreds of pages to cover grammar, page layout, the forms of various types of technical documents (letters, memos, reports, abstracts, papers, book chapters, laboratory notebooks), conventions for composing visual aids (figures, tables, graphs, charts), avoidance of plagiarism, correspondence with editors and publishers, and so on. We haven't tried to write such a book. In fact, *nothing* about this book is comprehensive.

We have two goals for this book. The first is to convince you that *to write like an engineer, you must think like an engineer*. The second is to provide you with a book that is easy, fast reading and strikes at the core of what you should know about engineering writing in order to do a good job. We're sure you're busy and we recognize that technical writing may not be your top priority, but we believe that there is a certain irreducible set of tools you can quickly learn to become an effective writer. Their use can save you time, energy, and frustration. They could also help you get some real appreciation for your technical ideas.

One of the ironies you'll notice about this book is that it centers on formal documents but is not itself a formal document. The chatty style we adopted (talking about *we* and *you*, for instance, and using contractions such as *don't* and *shouldn't*) was intentional. We wanted the book to be friendly and rapidly readable. For actual examples of formal writing, please focus on the items labeled **Example**. Some of these are good examples and some are poor (in our opinion), but at least they lie in the realm of formal, technical writing which constitutes the objective of the book.

With these things in mind, let us begin.

1

Introduction

1.1 Why Bother?

So you're an engineer or a student preparing to enter the profession. In any case, you're already busy. Your first question could be whether learning to write well is *really* worth much care and effort. We think it is. After all, an engineer who writes poorly might just

1. struggle and waste time;

2. disappoint or annoy important people like professors, mentors, supervisors, colleagues, and customers;

3. fail to complete a crucial document such as a thesis, dissertation, technical report, or technical proposal;

4. fail to get hard-won technical ideas across to others;

5. suffer from career stagnation or failure to land an attractive job;

6. alienate customers or lose contracts;

7. become ensnared in a lawsuit; and possibly

8. acquire a negative reputation as a poor communicator.

In short, a poor writer may have a second-rate engineering career. An ability to write professionally is a required part of being professional.

1.2 Think, Then Write, Like an Engineer

As engineers, we write primarily to communicate our thoughts rather than, say, our emotions or poetic sense. This is not to say that technical writing must be robotic or sterile. Like completely dry meat, completely dry text can be hard to chew. It's great if passion for a subject comes through in writing. But the *primary* goal of technical writing is usually not the communication of emotion. We want to tell our readers *what, where, when, why,* and *how.* So our writing should represent our thought. Like it or not, people judge us by the quality of our writing. We cannot be everywhere, but our writing can be (and, as a result of the Internet, likely *will* be). It could easily outlive us, acting as a helpful resource for — or a perplexing annoyance to — subsequent generations of engineers. We want our writing to be good.

What counts as "good" writing in an engineering context? This question will be addressed throughout the book. It is worth stating at the outset, however, that we will keep the needs and purposes of the *reader* firmly in mind at all times. We will have to consider the reader: his or her background, goals and purposes, etc. We are writing to inform a certain target audience, not simply to fill pages with words, equations, and diagrams. On the other hand, we are not aiming for "perfection" (whatever that is). We are simply trying to do a job with our writing and to do it well. Standard engineering design processes do not produce perfect switches or pulleys, and sound writing processes do not produce perfect documents. Intelligent compromise is basic to all practical engineering activity.

Engineers exist under constant pressure to be productive. They're expected to tackle large tasks, often with little guidance. A substantial writing project can be such a task.

It's time for Ken, a graduate student in mechanical engineering, to start writing his graduate thesis. Ken has a solid understanding of his research but is unsure how to plan a document of this magnitude. As the weeks pass, Ken begins to worry that he has writer's block.

Suppose we face a hefty writing project such as a doctoral dissertation, major technical report, or research proposal. How do we get started and stay on track? This question will also be addressed throughout the book. The good news is that writing can come naturally to engineers if they keep their task in an appropriate mental framework. We have found the standard engineering design process to be an excellent framework to use.

Dr. Smythe, Ken's thesis advisor, reminds Ken that he's an engineer and that time is money. The grant under which he's being supported has deadlines for certain deliverables, and one of those is a report based on Ken's thesis. Writer's block is not an option. The professor also reminds Ken that the very word *engineer* is derived from the Latin word for *ingenuity.* This switches Ken into more of an engineering mode: "I have a big problem to solve here. That problem involves writing, but it's still a problem and engineers are trained problem solvers. What do I know about problem solving, in general?" His next thought is "How could I *design* a document to solve my problem?"

Ken is on the verge of a realization. By using familiar engineering design concepts, he *can* tackle a huge writing project and do it well. First, however, Ken must adapt what he knows about design to the writing task.

1.3 Quick Review of Some Design Concepts

The engineering design process is often depicted as a flowchart. Figure 1.1 shows the main stages in the process. They are

1. Understand the goal.

2. Do your research. Gather data.

3. Generate solution alternatives.

4. Choose a solution to pursue.

5. Implement your solution and evaluate it. Improve it if necessary.

These steps can be adapted to virtually any formal technical writing task.

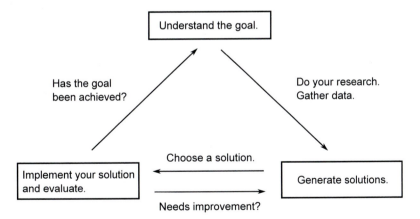

FIGURE 1.1
One visual representation of the typical engineering design process.

After reviewing the design concepts cited above, Ken responds by drafting a plan as follows.

1. Understand the goal. Ken must produce a graduate thesis that is acceptable to his advisor and his examination committee.

2. Do your research. Gather data. Ken must clearly understand the problem he faces. He must learn what a graduate thesis is. Who is the typical audience for a thesis? Is there a deadline for his final submission and defense? Ken must put all of his materials in order. These include the results of his initial literature search, his theoretical work and predictions, descriptions of his experimental setups and preliminary results, and his tentative conclusions. He knows that his research may not be complete; further experiments are needed, and the results of these may necessitate more theoretical work. However, he was told to begin thinking about the thesis, so he responds by planning to gather what he currently has available. Realizing that there are probably format requirements for something as important as a graduate thesis, Ken decides to seek a list of official guidelines from his university.

3. Generate solutions. Ken will have to develop a set of alternative approaches to writing his thesis. At this point, a major question is how the thesis should be organized at the chapter level. Rather than planning to generate just one possible response to this question, Ken follows the standard design process and plans to generate *two* possible chapter schemes. He hopes a discussion with his thesis advisor will assist him in choosing one of these schemes, combining the strengths of both schemes into a new scheme, or abandoning both schemes in favor of a third scheme he hadn't thought of yet.

4. Choose one solution to pursue. For Ken, this choice will be heavily influenced by discussion with his advisor — a primary "customer" his final "design" must satisfy.

5. Implement the chosen solution, evaluate it, and refine it if necessary. Ken understands that this is where much of his work will lie. He will have to write, solicit feedback from others, evaluate their comments, rewrite, try to obtain feedback again, and so on.

By adapting Figure 1.1 to his purposes, Ken has progressed far beyond merely getting started — he has generated a basic but essentially complete roadmap for himself. By following this map, he will avoid getting lost somewhere in the daunting process that stretches into the weeks and months ahead. Ken knows that at some point, if he persists, he will see the light at the end of the tunnel. It will be time to put the finishing touches on his document and arrange an opportunity to defend its contents before the graduate committee.

As an engineer, you can adapt your writing tasks along these lines. Whether you're writing a proposal, a report, or even a doctoral dissertation, a design-based approach will help you start a project, write more efficiently, and finish the task more quickly and with better results. Let's delve more deeply into some of the fundamental aspects of the design process.

Top-Down and Bottom-Up Approaches

Given a design task, we may decompose it into subtasks and, in turn, further reduce the subtasks to any desired level of decomposition. This idea is the basis of the *top-down* design approach (Figure 1.2).

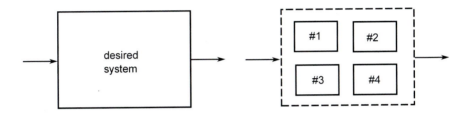

FIGURE 1.2
Notion of top-down design. *Left:* Top-level view of desired system. *Right:* Desired system is decomposed into a number of subsystems, labeled #1 through #4 here. Each subsystem may be further decomposed until the concrete design stages are reached. This is the analysis phase. In the synthesis phase, the completed subsystems are assembled into the desired system. The basic idea is divide and conquer.

One might take a desired electrical system and break it down into subsystems such as a microcontroller, a sensor, an actuator, and a power supply. The power supply, in turn, might be decomposed into a rectifier and a filter, and so on. Lower level subsystems are then designed and assembled into subsystems at the next higher level until the whole system is complete. This approach — analysis followed by synthesis — is strongly advocated as an engineering design approach and clearly applies to the design of large, formal documents.

The heart of Ken's research is a complicated experimental procedure that took his research group several years to perfect. Ken begins by writing several long paragraphs to try to describe the procedure, but soon realizes the difficulty a reader will have absorbing all this via one continuous explanation. He knows it will be easier to explain this procedure if he breaks it down into the sequence of steps required to actually perform the experiment. Ken decides to arrange his chapter on experimental approach so that each step in the procedure becomes a section. This lets him elaborate sufficiently without losing the reader in a multitude of details.

In contrast, we might imagine a research and development engineer who is free to pursue his technical whims every day. He generates a novel idea and implements it as an electronic circuit with no definite application in mind. A week later he thinks of a small upgrade to this circuit. Seeing the resulting improvement, he decides to feed the circuit output into another device he designed previously. The two subsystems show a curious interaction, and this gives him further ideas. A year later, he has a nifty prototype for a potential consumer product that nobody knew they might need or want. The design process described here does not fit the top-down category but is a design approach nonetheless. While *bottom-up* design is less advocated as a formal approach than the top-down paradigm discussed above, much design actually follows this pattern. Some people, including writers, prefer to work in bottom-up mode. When you find yourself unable to start or even plan a major writing project, try shifting into low gear and dumping your material into an electronic document so you can play with it. Move it around, reword a few things, insert dummy placeholders, etc., until organization begins to emerge. Small successes can lead to larger successes, and pretty soon a sensible document could evolve.

Ken plans to relegate lengthy derivations to an appendix of his thesis. He wants to concentrate on his main text, but also wants to collect appendix-related material as he goes. Ken does this, moving material into a separate computer file that will later become his appendix. Several weeks later, Ken has accumulated a file full of miscellaneous derivations. He can then organize the appendix using a top-down approach.

Some design processes can be viewed as *hybrid* approaches with both top-down and bottom-up elements. Engineering is about what works. We should state, however, that a key aspect of both approaches is *iterative improvement*. Iterative means repetitive; it refers to an approach where successive attempts are made, each one building on the previous one. You start with your chosen solution (recall Figure 1.1) and take successive cuts at improving it until it meets the standards for completion. Engineering design and the writing of engineering documents are the same in this respect. An iterative improvement loop is shown in Figure 1.3. In writing, iterative improvement is called *revision*.

Notion of Concurrent Design

Imagine a group of engineers thinking hard about a new project. They take several months to formulate a design proposal, only to have it shot down because they failed to consider post-consumer disposal problems and other

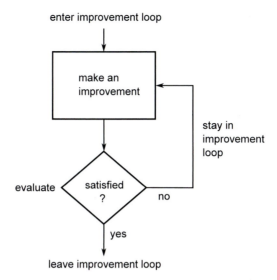

FIGURE 1.3
Flowchart segment for an iterative improvement loop. In technical writing the "evaluate" step might be labeled "proofread."

"downstream" issues. That's the risk we run with a *sequential design* approach. Sequential design means that design engineers work first in isolation, only later getting reactions from the other professionals who will eventually have a say regarding (and, possibly, veto power over) the project. What if that group of engineers had started by forming an integrated design team including experts in marketing, manufacturing and assembly, service, recycling and disposal, and so on? Such an approach is often called *concurrent design*. The two approaches are contrasted in Figure 1.4.

As a writer, you probably will not have the luxury of assembling a team of experts to help with a document. Nonetheless, the basic idea of getting information and early feedback certainly applies. If you plan to write a journal paper, for example, it may be wise to pick a target journal before starting to write. Engineering journals differ widely in their orientations and formatting requirements. A given publication may require everything to be written in the third person. Early awareness of such factors can save much work. You may also seek feedback early and often, providing drafts of a document to someone in (or close to) the target audience and asking for reactions. Their suggestions, if addressed early in the writing process, could prevent headaches later on.

Having completed and proofread a first draft of his chapter on the complex experiment, Ken gets feedback from the other students in his group. Fortunately, one of the students who developed part of Ken's experimental

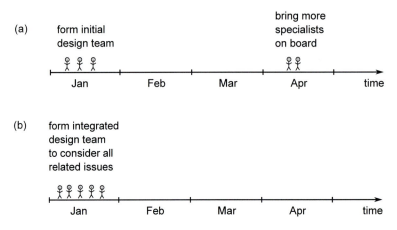

FIGURE 1.4
Sequential design vs. concurrent design: (a) sequential approach, (b) concurrent approach. The concurrent approach requires more initial structure but can prevent downstream conflicts.

procedure manages to catch an error in Ken's explanation. Ken corrects the error and thinks, "Thank goodness I fixed that before sending the material to Dr. Smythe for review." Ken submits the second draft and receives plaudits for such a well-organized chapter.

1.4   Chapter Recap

1. Careless or incompetent writers take needless career risks and limit their professional potential.

2. As engineers, we primarily communicate *what*, *where*, *when*, and *why*. Sometimes *who* is important as well.

3. Good technical writing is accurate and appropriate for a particular target audience.

4. The generic engineering design process applies to the design of formal documents.

5. You can attack a writing project by the divide and conquer (top-down) approach, the incremental accumulation (bottom-up) approach, or a hybrid approach. These approaches are customary in engineering design. Iterative improvement is essential to any approach.

6. Gather pertinent information on the target venue before starting to write. Learn the rules and guidelines.

7. Get feedback early and often. (And, as a professional courtesy, provide feedback to others who seek help with their writing projects.)

8. Thinking like an engineer is not just a paradigm for writing; it is also a framework for evaluating the written work of others. Engineers must review and critique many types of documents including journal papers, theses, proposals (external or internal), technical reports, white papers, and pitch slides.

1.5 Exercises

1.1. What types of design constraints do engineers routinely face? List as many as you can.

1.2. Pose a simple problem and generate at least three alternative solutions.

1.3. Write a short technical paragraph and have an associate read it. Ask what he or she got from reading it. Use this feedback to revise your paragraph.

1.4. Todd has reached a frustration point with his master's thesis. He brought 50 pages of written material to his thesis advisor who sharply rejected it for a number of reasons. Some pertained to specific technical aspects of the research, some to formatting issues of which Todd was not previously aware, and some to how the existing material had been organized into chapters. Todd is disoriented and ready to give up. What advice would you give him?

1.5. Choose a topic within your area of knowledge and consider how you'd explain it to a layperson. How would you break it into manageable chunks? Would you have to further decompose some of these chunks to make them understandable to the reader?

1.6. Pick an engineering, scientific, or technical journal in your area of interest, and study its mission statement and formatting requirements.

1.7. Engineering programs in North America are accredited by the Accreditation Board for Engineering and Technology (ABET). What does ABET have to say about the importance of writing ability for engineering graduates?

1.8. How has globalization affected the need for good, clear writing?

1.9. Explore the ethical codes published by professional societies such as the IEEE and ASME. Do any of the provisions of these codes carry implications about the quality of an engineer's written communications? Specify.

1.10. Name some general issues that are commonly addressed in technical writing. One example is *adequate functionality*: whether some entity will perform well enough in a given situation. The "entity" in question could be a device, a system, an algorithm, an approximation, an approach, or even a philosophy. For additional hints, see the chapter entitled "Analytical Reports" in the book *Technical Writing* by John M. Lannon (the full reference appears on p. 147).

2

Clearly Understand the Goal

2.1 What Is the Goal?

Whether your aim is to finish an advanced degree, meet a company deadline, or secure research funding, you will eventually have to compose a substantial, formal engineering document. Fortunately, the writing task can be made less daunting through an application of the engineering design process. Recall that the first step in that process (Figure 1.1) is to thoroughly understand your goal. As a writer, your fundamental goal is effective communication. Let us therefore state the goal as follows.

> Your goal as a technical author is to communicate information to an appropriate target audience.

In other words, some body of information currently resides in your mind and must be made available to the minds of the target readers. There are obviously two main issues here.

1. How does the information reside in your mind? How did it get there, and what forms does it take? Which portions of the body of information exist in your mind as visual impressions ("pictures"), abstract concepts, equations, quotations from authorities, etc?

2. Who is the target audience? What are the attributes common to those people at whom you will aim your finished document?

If your knowledge of a topic isn't clear to you, or if it isn't clear who you're trying to communicate with, your chances of true success with the writing endeavor will be small. Not convinced? Consider this analogy. An engineer accepts the task of designing an FM radio but fails to adequately consider (a) the parts available for building such a radio, or (b) the actual intended function of a radio (intercepting radio waves from broadcast stations and converting the information carried by those waves into human-audible form). Rather, his approach is to grab whatever parts might be at hand, solder them together without much of a plan, and see what happens. Surely his chances of coming up with a working radio are near zero. No properly trained engineer would approach technical design in such a way.

Unfortunately, many engineers seem to compose their formal documents in precisely that fashion. They effectively pile up what they hope will be "enough" pages of words, figures, equations, headings, stock phrases, etc., and send the result out the door to an unknown — or, at best, ill-considered — readership. Granted, this may seem an apt response to a truly pressing issue, such as

How can I meet this particular deadline and therefore keep my job?

But it is an untrained response, and we should not confuse "pressing personal issue" with "engineering goal statement." To make things worse, poorly conceived goals can cause problems. When companies rush flawed automotive designs to market, we see massive and costly product recalls. When graduate students submit unreadable theses, they're often sent back to square one. Consider the resulting cost in terms of lost time and energy.

Example. Peter refers to himself as a "graduating senior" but is still eight credits away from his bachelor's degree in civil engineering. It remains for him to complete his capstone design courses. Peter received a grade of "incomplete" in Capstone Design 1 because his final design report was deemed substandard by a faculty committee. So Peter cannot register for Capstone Design 2 and must spend much of the next semester revising his report. This could delay his graduation by as much as a year.

In successful technical writing, as in successful design engineering, *form must match intended function.* That all-important match will not likely happen automatically or at random. Remember the goal statement:

Your goal as a technical author is to communicate information to an appropriate target audience.

In order to communicate something of technical value to others, you must be clear about what you know and how you know it, understand the characteristics of the intended audience, and consider how to best transfer your knowledge to that specific group of readers. Let's examine the first two of these aspects in greater detail.

2.2   How the Information Resides in Your Mind

Engineering information takes many forms. A given fact may reside in your mind in the form of a visual image such as a specific electrical schematic or a mental impression of intermeshing gears. It may be an abstract concept such as "thermal energy" or "entropy." It may take a symbolic mathematical form such as the equation $V = IR$ for Ohm's law, or Maxwell's equation

$$\oint_C \mathbf{H} \cdot \mathbf{dl} = \int_S \mathbf{J} \cdot \mathbf{dS} + \int_S \frac{\partial \mathbf{D}}{\partial t} \cdot \mathbf{dS}$$

expressing Ampere's law. It may even take the form of "just words," although caution in such cases is warranted (remember the admonition of your college engineering instructors, that engineering is about thinking rather than memorization).

It is not our desire to embark on an adventure in psychology, delving into how information might be stored in the human brain. But we cannot afford to be unaware of how we grasp the essential aspects of our own writing topic if we hope to make a meaningful connection with our target reader. So we merely suggest that you, as an author, try to maintain some awareness of the issue while writing.

Example. Jim is writing a technical report on his latest project. His view of the final design — for a cell phone circuit — is mainly a visual one: when he thinks about the circuit, it seems natural to do so in terms of a picture. Jim therefore decides to introduce his readers to the electrical layout by way of a photograph paired with a labeled line drawing. The drawing will call attention to certain key features of the photograph (remember, the camera captures *everything*, whether conceptually important for present purposes or not). After all, it took him months of work to formulate this layout; could a reader really grasp it immediately from a photograph alone? A good drawing will take effort, and Jim entertains the possibility of omitting it. But he follows his better judgment and painstakingly constructs a line drawing. Several years later, while referring to the resulting report as a memory refresher, Jim is pleasantly surprised when that same line drawing guides him smoothly through what now seems like a complicated photograph. Finding himself unexpectedly in the position of a target reader for his own report, he discovers that the document has *archival value*.

If you're the subject-matter expert and *you* find it helpful to understand some particular aspect of your topic in a particular mode, then why not give your reader the same opportunity? If you think it's best understood in equation form, then present it that way (carefully, of course).

Example. A table can be a clear way to present things. Consider the following table, exhibiting the analogous properties of electrical capacitors and inductors:

property	capacitor	inductor
construction	plates	coil
dominant field	electric	magnetic
circuit parameter	C (farad)	L (henry)
terminal relation	$i = C\,dv/dt$	$v = L\,di/dt$
integral form	$v = \frac{1}{C}\int i\,dt$	$i = \frac{1}{L}\int v\,dt$
dc behavior	open	short
energy stored	$\frac{1}{2}Cv^2$	$\frac{1}{2}Li^2$
continuous variable	voltage	current
series	add as reciprocals	add
parallel	add	add as reciprocals

Would these parallels be as clear from a text-based presentation?

2.3 Your Audience

Effective writers put their audiences first. As an engineer, you are writing to inform — not to bluff, dazzle, impress, or enchant, and certainly not to confuse or frustrate. Think about the potential reader: try to envision him or her.

1. What is the reader's background? Is he an accomplished expert with education specifically related to the topic, or simply a consumer who might have to refer to an owner's manual for your product? Is he a degreed electrical engineer, or a degreed mechanical engineer who needs to understand a circuit design anyway?

2. What are the reader's purposes? Will she simply want the bottom line regarding the topic? Will she require the level of understanding needed to replicate the work in detail?

3. What is the reader's likely level of understanding? Is he an undergraduate student? Is he the highly educated target reader of a typical scientific journal?

Pose these sorts of questions and try to answer them before proceeding with a project. The reader is central to all considerations regarding a document. Writing a document without trying to understand the target audience is like designing a product without trying to understand the potential customer. Remember, it's always helpful to view your writing through the lens of engineering design.

Example. Rosa assembled an internal document describing how engineers should incorporate the newest computer modems into her company's products. Now she is troubled to hear complaining that her tutorial is hard to understand. Rosa's boss Pam suggests that the problem is with the organization of the document, which differs from that of the company's other tutorials. The difficulty, Pam claims, is a lack of familiarity by company engineers with the layout of the document. Perplexed, Rosa explains that this layout was the norm at her last company; she can't see why it shouldn't work here just as well. "Think of why we use IEEE 802 standards for our modems," Pam says. "What would happen if we tried to incorporate noncompliant modems into our products?" "That's obvious," Rosa replies, "the systems wouldn't know what information to expect, and couldn't communicate with each other." Pam points out that writing works the same way. The reader expects certain conventions to be followed in grammar, spelling, *and* layout. When conventions are disregarded, confusion results and the line of communication is disrupted.

The idea that writing conventions can be viewed as engineering standards made it easier for Rosa to plan and execute her next big writing project, avoiding the confusion experienced by the readers of her tutorial. Let's add a few similar analogies and summarize them in tabular form:

writing project	engineering design project
audience	customer
writing conventions	engineering standards
author time and effort	project cost
document brevity and conciseness	product efficiency
document clarity	product effectiveness
critique	customer feedback

It is clear that, aside from terminology, technical writing is not fundamentally distinct from the rest of engineering activity. Like any consumer product, your writing must address your customer's needs. It must adhere to expected conventions such as those of English grammar. It must economize the reader's time, energy, and patience. Finally, it must come in under budget in terms of your own time and energy as a writer and a busy engineer.

Notion of Communicative Accuracy

Let's say it again: good writing is that which is helpful to the *intended target audience*. Your document does *not* have to satisfy *every conceivable audience*. It would be difficult to compose a technical document that satisfied all readers from high school students to university researchers (the difference in mathematical backgrounds alone could make it impossible). So how should we proceed in cases where we must address an audience whose preparation is less than ideal? This question is routinely faced by science writers, and a classic response was put forth by Warren Weaver in an editorial entitled "Communicative Accuracy" in the journal *Science*.[1]

Weaver was addressing scientists who, for whatever reason, had to describe their activities and discoveries to lay audiences. His two *conditions for communicative accuracy* were as follows (we paraphrase them for engineers):

1. Your statements should meet the reader at his current level of understanding and move him toward a better level of understanding.

2. Your statements should not mislead the reader (e.g., through too much simplification) in such a way that an enhanced level of understanding will be blocked later on.

Whether or not you choose to be guided by these principles, please note that they center on the needs of the reader — not on those of the author.

Example. Throughout the book, we will emphasize the helpful role of appropriate analogies in explaining certain technical matters to a reader. Laura, a chemical engineer, must pitch a technical proposal to a mixed audience. Although the audience includes some highly informed engineers, the principal decision makers for proposal funding will be the relatively nontechnical owners of the company where Laura works. (The company has been family owned and operated for several generations.) In her brief description of atomic structure, Laura says

> Let us recall that the atom is somewhat like a tiny solar system, with the nucleus occupying the position of the sun and the electrons orbiting the nucleus like tiny planets.

Is this a rigorously correct picture of the atom as understood by the modern physicist? Certainly not. However, *relative to her target audience*, Laura's description of the atom *does have communicative accuracy* in Weaver's sense. Let's check Weaver's two conditions:

1. Laura's statement does give business managers some picture of an

---

[1] Volume 127, Number 3297, 7 March 1958, published by the American Association for the Advancement of Science.

atom (a picture they once had in high school but since forgot). The picture is workable for purposes of comprehending the proposal as a financial decision maker. In the present situation, Laura decided that a solar-system picture would beat, say, a picture of an atom as a microscopic animal or a tiny steam engine. She got them closer than their previous level of understanding, which was nil.

2. Her careful use of the words "somewhat like" kept the door open to better understanding, in case one of her audience members becomes intrigued and feels like opening a book on atomic structure someday. She did *not* say that the atom "is a tiny solar system with little balls whizzing around a heavier center ball called the nucleus." This is a very different statement and it could really block the reader from gaining a truer understanding of the atom if he chooses. By attaching too strongly to the picture of an electron as a particle, the reader could have a hard time accepting the electron's wave-particle duality, for example.

Weaver's two conditions are therefore met.

Handling the Ill-Defined Reader

What about that rare situation in which the traits of the potential reader are really hard to discern? We can still offer some general suggestions.

1. Respect the reader; give him or her the benefit of the doubt. You do not wish to insult the reader with technical baby talk. On the other hand, you cannot assume he or she already knows your subject area. (Otherwise why is he or she in the target audience?) Therefore, a good rule is to assume that the reader will be slightly more intelligent than you are, but much less knowledgeable about the subject area.

2. Rely on formality, convention, and consistency. Although these aspects of traditional writing are too often ignored in this era of free expression, they still represent your best shot at establishing communication with a totally unknown reader.

Example. Ahmed is writing a user's manual for a piece of equipment. The gadget could be used by virtually anyone, at all levels of education and across multiple cultures. In order to decrease the possibility of confusion, Ahmed decides to implement a high degree of visual structure in his document, peppering it with "signpost headings" such as *Caution, Comment, Example, Guideline, Limitation, Procedure, Summary,* and *Warning.* Of course, someone somewhere might not get the appropriate messages from these words, but they are common English words whose meanings are conventionally understood. Ahmed is wise to use as many of them as possible in his present circumstance. He wants his warnings to be taken as *warnings*, not as suggestions or optional side notes.

One engineering analogy for convention and consistency in writing is the familiar notion of *user interface*. Good user interfaces are intuitive. It makes no sense to have

- a control box with six switches, where the up position means ON for five of them but OFF for the sixth one, or

- a machine with the emergency stop as a small green button, or

- knobs that increase a quantity (such as volume) when rotated counterclockwise, or

- a computer application that asks the user to enter data sometimes in pop-up boxes, sometimes in the bottom corner of the window.

Even if you have no idea who will run your machine, big and red is a better choice for an emergency stop button than small and green. People, in general, expect some convention.

The same thing holds in technical writing. Stylistic conventions are worth understanding. Confining but helpful, they can save time in the long run and make you more understandable to others. We will have much more to say about specific writing conventions in the following chapters.

2.4 Other Aspects of Situational Awareness

You may have to keep other things in mind besides the reader. Here are a few questions to ask during the planning stages.

1. Does this project have a time limit? If so, is the deadline self imposed or externally imposed?

2. Will this be a single-source document? Will you write the entire document alone, or will it be a collaborative effort?

3. Should this document have archival value? Is the document intended to have long term value, or is it intended for short term use only? Are you *really* writing a thesis *just* to get your degree? A quality thesis could be an asset later, as a technical reference or tool for use in your job search. You might wish to have a thesis document that provides a sense of pride in a job well done.

4. What are my available resources? In addition to time, you may have to consider such things as working environment and information access.

Example. Sandra has agreed to act as editor of a new engineering handbook. Her publisher has put her in charge of soliciting contributions from experts worldwide. Most of Sandra's correspondence will necessarily be through electronic mail. Although Sandra is exquisitely organized herself, she overestimates the time-management capabilities of all her contributors. She excitedly signs a publishing contract, promising to have the final manuscript finished by year's end. Sandra is stunned when in late December she finds herself waiting for over half of the individual chapter manuscripts to arrive.

2.5 If Persuasion Is Part of the Picture ...

This book deals primarily with factual writing. We do recognize, however, that a technical idea typically must be pitched and sold before it can get attention or funding. We can easily envision (if we haven't already experienced) the peer reviewer — let's call him Fred — who rejects a sound paper or proposal simply because it is "not sufficiently well motivated." Months spent on a document are of little use if Fred won't give it a serious look. Worse yet, the document may sit on Fred's desk for half a year before he gets around to stamping it with his famous[2]

> Idea seems sound, but no apparent application.
> *Recommendation: Reject.*

Assuming our document *is* technically sound, how can we give it a better chance in the competitive marketplace?

The effective use of English falls under the heading of *rhetoric*, and many books on persuasive writing are available. As engineers are bound by ethical codes, we will not explore the tricks used by salespersons and other persuasion specialists here. Nonetheless, we offer a few suggestions that you may have occasion to work into a document.

1. You must quickly gain the reader's interest. Otherwise he may stop reading and direct his time and energy elsewhere. Assume you have an extremely narrow window of opportunity. This brings us to ...

---

[2]We shouldn't be too hard on Fred. Peer reviewing is usually done on a volunteer basis. Moreover, Fred's papers and proposals are treated the same way by his own resource-conscious peers.

2. You must somehow address a need for the product. The product could be your document or an idea it contains. Perhaps you're writing to inform people who need to know about a subject. Maybe you're writing to advance an idea that will solve a significant technical problem. The point is that you must communicate these things, possibly to a skeptical and distracted reader. Don't expect a complex technical document to act as evidence of a need for its own existence.

3. You must offer a solution. You must argue (or at least state) that the product addresses the established need.

4. You may have to differentiate your product from its competitors. That is, you may have to argue that the product addresses the need better than other available products.

For the reader, these things may constitute the only available reasons why your product should be seen as useful and worthy of consideration.

Example. Bob is a civil engineer who runs a small construction company. He has been unable to get a contract with the county government, since a few large regional companies dominate the local construction market. Bob wrote several proposals emphasizing his success in various small projects, but this was never enough. Perplexed, he asks a former college roommate Jack, who runs a similar company, to read one of his proposals. Jack notes that the regional companies are also successful, and Bob has failed to provide a good reason why the county should choose him. "What makes your company special?" When Bob points out that his company has a program to hire returning military veterans, Jack suggests he build his proposal around this community service practice. The county is persuaded and awards Bob a long-term infrastructure contract.

2.6   Chapter Recap

1. As engineers, we understand that it's hard to solve a problem without understanding it first.

2. Your goal as a technical author is to communicate information to an appropriate target audience. In other words, a document is intended to communicate what you know to an interested, reasonably prepared reader.

3. Producing a formal document *just* to meet a deadline or attain some other reward is subject to the old rule of *garbage in, garbage out.*

4. It helps to think about what you know, and how you know it, before trying to write about it for someone else. Doing so might even help you become a better subject-matter expert.

5. A picture may be worth a thousand words, but a photograph may be worth a million words and the reader may not have the time, energy, or motivation to digest that much. A nicely labeled and explained line drawing can be priceless.

6. It's essential to consider your target reader's background, purposes, and maturity level. This is especially the case with mathematical maturity.

7. If you dislike writing but enjoy pleasing your customers as an engineer, then think of your reader as your customer.

8. The principle of communicative accuracy offers guidelines for communicating with lay readers.

9. If you can't picture your target reader in any other way, then at least picture an intelligent, busy person.

10. It's hard to go wrong by leaning on formality, convention, and consistency.

11. Big writing projects are complicated. You must be aware of many factors. The best approach is to become aware of them as early as possible.

12. Engineers are often called upon to persuade. People are busy; they may not give your writing a chance if not provided with an adequate reason.

2.7 Exercises

2.1. List some broad purposes for technical writing. For example, some documents are intended to provide instruction whereas others are merely intended to record information.

2.2. Pick a published paper in an engineering journal whose title interests you and read one full page. Try to form a picture of the target audience (it may not be easy, but try). Repeat with one page from an article in a technical trade magazine.

2.3. If an idea dawned on you in a flash of insight, must you still lay it out systematically for the reader? Why or why not?

2.4. Make a list of signpost words that could serve as headings or subheadings in your own writing projects.

2.5. Evaluate the assertion

Engineering is about building things!

for communicative accuracy with respect to the following audiences: (a) a kinder-garten class, (b) a high-school science class, (c) the parents of prospective college students, and (d) engineers gathered at a technical conference. Repeat for the assertions

Engineering is the application of scientific and mathematical principles to solve real-world problems.

Engineering is what engineers do.

2.6. Criticize the following passage:

Therefore, $F = 6$ N. Plugging this into our previous result, we obtain $a = 5$ m/s^2. Note that the last term in equation (5) went away in this case.

2.7. Write some brief instructions on how to use a hammer. Aim your piece at a seven-year-old child, emphasizing function, appropriate usage, and safety.

2.8. Organize some technical information in table form. Choose any topic of interest.

3

Mindset for Technical Writing

We have encouraged you to think about your technical writing project as a problem to be solved, and then use the engineering design process to attack the problem. For the remainder of the book we emphasize the crucial portion of the flowchart in Figure 1.1: *implement, evaluate, improve.* Let's start by considering some tools you can use.

3.1 See Rules as Helpful Tools

Rules are not always correct *per se*; they sometimes require modification when times change. But broadly accepted rules did evolve for reasons. They can help us stay within the general bounds of convention and thereby increase the number of readers we can reach. They can provide us with structure and save us from having to think too much. We ignore rules about writing, especially active conventional rules, at our own peril.

Let's divide our rules into three categories: (1) rules of logic; (2) rules for English grammar and style; (3) other rules and conventions, including those for visual layout and the reinforcement of certain broadly accepted professional habits. These will be addressed in subsequent chapters. For now, we emphasize that rules are here to be helpful; try to think of them as parallel to engineering standards (recall the table on p. 15). With that accomplished, we proceed to a crucial aspect of mindset.

3.2 Think Clearly Before Starting to Write

There are times when engineers are called upon to describe objects, devices, or systems, and times when they are called upon to put forth logical arguments. It's important to understand that description and argumentation are not the same thing. Something to ask, before putting technical thoughts into written form, is[1]

> Should I be *describing* something right now, or *arguing in favor of* (or possibly *arguing against*) something?

Suppose, for example, you are presenting an electric circuit design. A full presentation may require both description and argumentation. However, you definitely want to be clear as to which expository mode you are pursuing at any given moment. You could, say,

1. start with a pure description of what the circuit *is*, then, after finishing the description, provide arguments for why the circuit should be (or had to be) that way, or

2. depart from a set of accepted definitions, physical principles, design constraints, etc, and display a tight logical argument leading to the circuit and its final description.

Either of these approaches is sensible. It's *not* sensible to launch into a description of something, then lapse into a loose display of logical-sounding fragments leading nowhere, then lapse back into description, and so on.

Example. The following passage is clearly descriptive.

> The system under consideration is shown in Figure 3.8. The input signal $x(t)$, an analog signal, is fed into an analog-to-digital converter, the output of which goes to port A of the microcontroller. The software algorithm evaluates the information and makes decisions on where to position the robot arm.

This passage tells us *what is*, not *why it had to be that way*. It is not argumentative. The author does not draw conclusions from given facts (deductive argument) or try to generalize from known facts (inductive argument). He or she concentrates, at least for the moment, on painting a picture.

---

[1]We depart from the four traditional categories of narration, description, argumentation, and exposition. Much of formal, technical writing is what an English teacher might call exposition. To us, *argumentation* does not primarily mean *persuasion*. We're engineers writing for engineers. Our main objective is not to study writing; it is to promote a mindset conducive to clear writing.

Example. This passage is clearly argumentative:

Let R_1 and R_2 be two resistances connected as shown in Figure 7.5. The relation between the output voltage V_0 and the input voltage V_i is given, according to circuit theory, by

$$V_0 = V_i R_2 / (R_2 + R_1) \, .$$

Since we require $V_0 = V_i/2$, we can make any choice of resistors satisfying
$$R_2 / (R_2 + R_1) = 1/2 \, .$$

Our choice is $R_1 = R_2 = 10$ kΩ. The resulting design is displayed in Figure 7.6.

Example. Consider this passage:

We now describe the system under consideration (Figure 4.1). Because the filter blocks labeled A and B are different, they do not produce comparable outputs. Furthermore, in view of the fact that an output between 4 and 10 Volts was required for stage C, the values of R_1 and R_2 are identical. Note that the output of stage F is again filtered before being made available to stage G. It is required that the final output (from stage G) be clean.

Is this description or logical argument? We say *neither*. It's merely an example of *bad* engineering writing.

What sorts of cues indicate whether an author is trying to argue rather than describe in a particular passage? The structure of logical argument is carried by

1. *definitions* ("Suppose we let F be the force acting on ...")

2. *logical implications* ("if ... then ...") and

3. *logical equivalences* ("... if and only if ...").

You will see *premise indicators* such as *since*, *because*, and *in view of the fact that*, and *conclusion indicators* such as *therefore*, *hence*, and *we conclude that*. Learn to recognize and use these appropriately in writing. The preceding example shows that their inappropriate use yields a terrible muddle.

We'll cover some essential points of logic, including the notions of definition, logical implication, and logical equivalence, in Chapters 4 and 7. Some pointers on description appear on p. 89.

3.3 Again, Keep Your Reader in View!

As we begin to present some particulars about writing, let's not forget about the reader. He or she must be an ever-present part of the writer's mindset. What is the reader's job in all this? The answer is *comprehension*: he or she can be expected to focus on your writing, trying like crazy to construct accurate mental pictures of the author's intent. *Your* job is to make that process easy, or at least possible! If you make it possible, your writing will be useful. If you make it easy, your writing will be useful *and* appreciated. Why not shoot for both? We'll offer a hint that could help with that.

Start and End Transitions: The Keyhole Model for Friendly Composition

Avid readers of popular scientific articles may notice a pattern. A typical piece will contain some real scientific meat, but the author won't start or end the piece with that material. Indeed, the audience members will be assumed to approach the piece from many different directions and initial mindsets. Some seek specific information, but others read out of plain curiosity, to relax, or just to kill some time before a dental appointment. Does the author want to give all these readers a chance? Definitely! So, does the following passage sound like a good way to start a popular magazine article?

The Magic of Amphibians

To really understand frogs and other amphibians, we must ask how they can regulate the water content of their bodies — both on land and in the water.

Sure, it's not all that technical, but would it appeal to a wide audience? Would it provide welcome distraction from a dentist appointment? The problem is that not all readers of such an article will want to *really understand* frogs; that's not their *initial* mindset, anyway. They might shudder and skip to the next article, hoping for a friendlier beginning, something that grabs them or at least provides a more gentle psychological transition. We'd better try again:

The Magic of Amphibians

Wow, that was a close one! A single step in the wrong direction in *this* marsh could plunge you into tea-colored water way over your

head. Yet we were after a truly worthy prize — a glimpse, in its
natural habitat, of the GooColored Tiger Frog. Suddenly, ...

Would you read at least one more word to find out what happened suddenly?
If so, you might even reach the core of the article where some discussion of
water regulation in amphibians (a true aspect of their biological magic as
multi-habitat creatures) would be better placed.

Does it take sensationalism or other cheap literary devices to "grab" a
casual reader nowadays? Maybe. But engineers writing formal, technical material must stay within professional bounds. Nor can they write about fas-
cinating topics like frogs in swamps. Nonetheless, there's a useful principle
here: if the opportunity presents itself (which it would not in a scientific journal article but *might* in a document for a broader audience), you could start
your piece with a recognition that the readers will probably be arriving from
somewhat different directions.

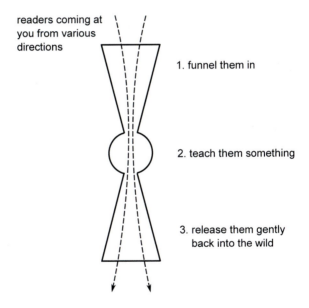

FIGURE 3.1
Keyhole model for friendly writing.

Imagine an old-fashioned keyhole (the kind that accepted skeleton keys in
Victorian times; see Figure 3.1). The top section is broad and could be used
to funnel a variety of readers into your piece. The meat is at the center. After
doing its main job of informing, the piece funnels the readers back out with
a non-abrupt end transition (in other words, it doesn't terminate with a final
fact about the subject).

The keyhole model is an example of conventional structure. It's not suitable for writing journal papers, because journals and their readers demand formal structures starting with abstracts and introductions and ending with conclusions and reference lists. But if you ever have to pitch a funding proposal to a nontechnical audience, or excite a group of lay people about what your company does, you could consider using the keyhole model. Look at it simply as another available tool that can keep you oriented toward the needs of your customer: the potential reader of your document.

3.4   Getting Started with a Mind Map

Some people like to start their writing projects with an outline. They arrange their ideas something like this:

Topic

1. First-level subtopic
    (a) Second-level subtopic
        i. Third-level subtopic
        ii. Third-level subtopic
    (b) Second-level subtopic
2. First-level subtopic
    (a) Second-level subtopic
    (b) Second-level subtopic
3. First-level subtopic
        $\vdots$

The formal outline is a classic way to begin; use it whenever helpful. At times, however, you may find it difficult to produce an outline. Perhaps the organization of the subject in your own mind is too unclear; perhaps it seems clear but hopelessly tangled. Maybe you know a lot about the topic but get stuck when trying to outline it. In such cases, you may find an alternative technique helpful.

The Mind Map for Writers

You could try the following. Take a large blank sheet of paper (or work with a computer drawing program), write the name of your topic near the center, and draw a dark (or double) oval around it. Now ask yourself what seem to be the main aspects of your topic. Each time you think of one, write its name inside an oval near the main oval and connect the two with a solid line. When you believe you have all of these first-level aspects, repeat to produce a set of what you consider to be second-level aspects (main aspects of the first-level aspects) — each inside an oval. A real creative flow can be encouraged in this manner: a mind map of your topic can spill onto the paper with seemingly little effort. The picture can literally bloom like algae.[2] The technique is indicated in Figure 3.2.

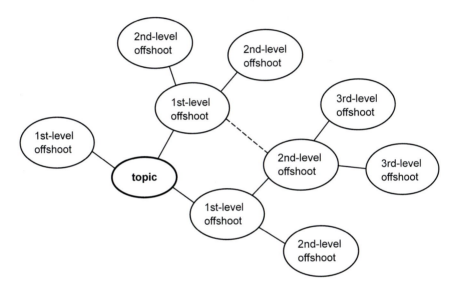

FIGURE 3.2
Notion of a mind map. The picture begins with the *topic* oval, after which the *offshoot* ovals suggest themselves.

During the process you may sense a need to indicate other connections between existing ovals. Do this with dashed lines. This can help you continue expanding your mind map even with subjects that are a bit tangled by nature. One such dashed line is shown in Figure 3.2.

Mind maps are interesting things. A mind map for a mature subject (i.e., one that has matured inside your head) will tend to be balanced in some way (at least you'll get that sense when looking at it). It may have a pleasing symmetry. If a mind map is asymmetric (like Figure 3.2) but feels complete, fine; maybe the topic itself is a bit lopsided by nature (there *are* topics like that). If it's asymmetric and feels incomplete somehow, it may be calling you to take a rest and continue. You can return to your mind map again and again until you get a sense of satisfaction from it. At that point you may feel

---

[2]Feel free to invent a different metaphor if you dislike algae.

prepared to (a) make a formal outline, (b) jump in and start writing, (c) set that mind map aside and try another one. If any of these things happen, the mind map has done its job.

Example of a Mind Map

When one author of the present book started to conceive the book, he made a mind map for Chapter 1. It is reproduced as Figure 3.3.

FIGURE 3.3
Mind map for an abandoned version of Chapter 1.

This mind map facilitated deeper discussion between the two coauthors: it effectively permitted one person to see, at a glance, how a topic (that of writing mindset) existed in the mind's eye of another person. After some discussion a crucial decision was made, and a new mind map was constructed for what would become the actual Chapter 1 of this book. It's just one example of the potential use of the technique.

3.5 Chapter Recap

1. There are tools to make formal technical writing easier and more effective. These include the rules of logic, the rules of English, and other rules we will examine.

2. It's hard to write effectively when in a poor mindset for the task.

3. Murky thinking leads to murky writing.

4. It's essential to be clear, at any moment while you're writing, whether you are describing something, constructing a logical argument, attempting to persuade as best you can under the circumstances, or something else. A lack of clarity about this will result in a muddle of words and equations. Your target reader may give up, and rightly so.

5. You can mix description and argumentation — even in the same paragraph if necessary — but you must be nimble and clear in your transitions, sending the reader the signals needed to follow your changes of direction.

6. The reader's job is to comprehend what is written — not to "mentally rewrite" in an attempt to compensate for a writer's carelessness.

7. So many technical documents are badly written that readers can hugely appreciate a skilled author.

8. Aim for smooth transitions in your writing, unless you wish to shock the reader (a risky practice at best).

9. The mind map technique is a way of coaxing information from your brain — sometimes tangled information. Mind maps can be fun and interesting as well as useful.

3.6 Exercises

3.1. Classify the following passage as description or argumentation.

Then, assuming we are far enough outside the structure, only the dominant field contribution will exist. Knowledge of $E(r)$ is sufficient to determine V at location P by virtue of equation (14). Hence, according to the concepts discussed in Section 1.5, the output I of the structure is fully determined.

3.2. Repeat for the following passage:

An example of such a structure was assembled from 39 evenly-spaced rods above a copper plate. Each rod was made of aluminum and had dimensions $a = 5$ mm and $l = 10$ cm. The entire structure was enclosed in high-grade plastic for weather protection.

3.3. Write a paragraph describing a technical item that interests you.

3.4. Write an argumentative paragraph, deducing some technical conclusion from a set of definitions and accepted premises.

3.5. Read one full page of any standard engineering textbook. Try to determine the author's basic intent with any given sentence. Is he or she trying to describe something, argue in favor of something, or seemingly neither?

3.6. Write a description of this geometric figure:

3.7. How would you start a popular magazine article about lightning?

3.8. The three classical elements of English composition are *unity*, *coherence*, and *emphasis*. Try to define these terms or describe their implications.

3.9. Construct a mind map for an interesting technical topic.

3.10. Construct a detailed outline for an interesting technical topic.

4

Avoid the Worst Thinking Traps

Engineering should be about doing things right. A system such as a bridge, telephone, or traffic light should be designed to function properly over a certain lifetime. It should satisfy its intended purpose. The same holds for a formal technical document meant to describe an engineering system, communicate an engineering procedure, or develop an engineering idea.

This book addresses the combination of thinking and writing skills necessary to produce effective formal documents. As engineers, we are not merely required to think; we must also avoid major pitfalls in our thinking. We ignore this responsibility at our own peril (and often at the peril of others).

Example. Tom maintained a high grade-point average in college by always demanding detailed instructions and following them to the letter. As a newly-hired chemical engineer, however, he is required to think on his own in virtually every situation. Tom is somewhat overconfident about the integrity of his thought processes. Today he must decide how a batch of hydrofluoric acid (HF) should be stored in the company's new lab room. He thinks to himself,

> Acids are stored in glass containers. HF is an acid, so it should go in a big glass jar.

Tom fails to verify that HF shares the properties of the other acids he's thinking of. In fact, while it's true that acids are commonly stored in glass containers, HF is used to etch glass. It obviously cannot be stored in a glass container. Tom causes a chemical spill in the company's new lab, and hazardous material and cleanup crews must be summoned at great expense. Tom learns that his high grade-point average holds little value in this situation; he soon finds himself seeking a new job.

Tom has committed a blunder called the Fallacy of Accident. We'll return to it later, but for now it's obvious that Tom should have done his homework rather than basing his decision on an assumption. Avoidance of common fallacious patterns is the main thrust of the present chapter.

4.1   Claims vs. Facts

In Chapter 2 we mentioned the role of persuasion in certain types of engineering documents. We cannot deny that a persuasive component is sometimes essential. A substantial portion of a patent application consists of claims that specify the essence of the invention and allow clear understanding of what will infringe the patent post issue. Although some portion of the claims consists of factual material, a main purpose is to persuade the examiner as to the novelty of the invention. Indeed, the purpose of even the most technical argument (such as a mathematical proof) is to persuade ourselves and others that the conclusion of the argument holds. For this reason, in this chapter we will emphasize the notion of *claim* instead of the notion of *fact*.

Many of the things written in technical documents are, from our present viewpoint, *technical claims*.

Example. These are technical claims:

All acids are dangerous.
The optimal value is $x_0 = 1.45$.
Some antennas are omnidirectional.

These are not technical claims:

Let $y_1 = 19$.
The purpose of this chapter is to introduce the gamma function.
We would like to thank Dr. S.P. Smith for his valuable guidance.

Indeed, it would be irrational to dispute any of them. The second, for instance, merely states an author's intention.

The attributes of a good technical claim include

1. clarity — the claim must be understandable;

2. verifiability — the claim must be supportable.

Clarity is greatly supported by grammatical structure, the topic of Chapter 5. For example, patent attorneys use a very specific syntax with strict grammatical rules to prevent the misunderstanding of claims. In this chapter we deal with issues related to verifiability. Along the way, we will touch on some elements of logic and heuristic reasoning. Our treatment is informal and we make no attempt to compete with textbooks on logic. Our main goal is to save you from writing things that are illogical or otherwise outrageous.

4.2 Logical Fallacies

First, we recall that *deductive argument* moves from accepted truths to their logical consequences. *Inductive argument*, on the other hand, attempts to move from specific observations to general consequences. These are the two basic modes of reasoning in Western thought. In this section we examine some elementary deductive fallacies that may appear in careless writing. Avoid them at all costs!

Syllogistic Fallacies

We begin with the famous argument

> All men are mortal.
> Socrates is a man.
> Therefore, Socrates is mortal.

This is an example of a *categorical syllogism*. It consists of two *premises* (or *assumptions*) and a *conclusion*. The argument is valid not because the conclusion holds, but because the truth of the conclusion must follow from that of the two premises. Any argument of the general form

> All S are P.
> x is an S.   ✓ valid
> Therefore, x is P.

is valid for the same reason. In Figure 4.1 we use diagrams to clarify the situation. Although these are not as rigorous as the *Euler* and *Venn diagrams* that appear in logic books, they will suffice for our purposes.

We must be on the lookout for the categorical syllogism. Such an argument may be shortened by leaving one of the premises unstated.

Example. Consider the syllogism

> All machines eventually fail.
> An automobile is a machine.
> Therefore, an automobile will eventually fail.

Leaving the second premise understood, we could write

> Machines eventually fail, hence an automobile will eventually fail.

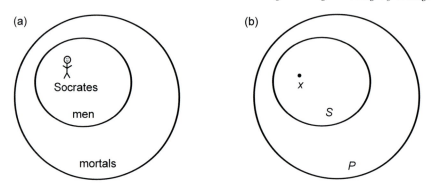

FIGURE 4.1
A categorical syllogism. (a) Concrete case. Socrates falls within the set of all men, and the set of all men is a subset of the set of all mortals. Hence Socrates must be a mortal. (b) Abstract case. The point x falls within the set S, which is itself a subset of a set P. Clearly x must fall within P.

Leaving the first premise understood, we would write

An automobile is a machine, so it will eventually fail.

Both of these are still considered syllogisms.

So far, so good. However, we must also guard against *syllogistic fallacies.* These arguments masquerade as valid syllogisms but fail the basic test mentioned above, i.e., that the truth of the conclusion *must* (not *may*) follow from the truth of the premises.

Example.

All bottles containing hydrochloric acid must be marked "Corrosive." This bottle is marked "Corrosive." It must contain hydrochloric acid.

The general form in this case is

All H is C.
y is C.   × invalid
Therefore, y is H.

The invalid nature of this argument is clear from Figure 4.2(a). The point y shown in the figure is an *invalidating counterexample*, and it takes only one of these to prove an argument invalid.

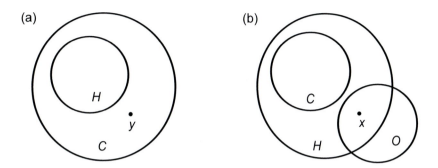

FIGURE 4.2

Two invalid syllogisms. (a) Everything that has the attribute H also has the attribute C. However, the additional assumption that y has the attribute C does not imply that it *must* have the attribute H. (b) The assumption that O and C are disjoint does not imply that O and H are disjoint. An element x can belong to both O and H.

Example.

> All chemicals in our lab are marked "Hazardous." No chemicals in our lab are organic. So no organic chemicals in our lab are marked "Hazardous."

The general form in this case is

> All C is H.
> No C is O.   × invalid
> Therefore, no O is H.

See Figure 4.2(b); the counterexample x shows that this argument is also fallacious.

These examples show that we must consider the very structure of an argument, not just the truth of the individual statements it contains. They also show how easy it is to fall into the trap of a fallacious argument.

In deductive logic, an argument is said to be *valid* if its conclusion must be true whenever its premises are true (or, as logicians say, the conjunction of the premises implies the inclusion). If, in addition, the premises are actually true, the argument is said to be *sound*.

Example. The argument

All engineers are wealthy. Jo is an engineer. Hence Jo is wealthy.

is valid. It is not sound, however, because the first premise is false.

Example. Here is a sound argument:

All positive real numbers are nonnegative. Ten is a positive real number. So the number ten is nonnegative.

The argument form is valid and both premises are true.

Needless to say, engineers should be advancing sound arguments.

Arguments Involving Conditional Statements

Before considering additional valid argument forms, we must modify our notation a bit. From now on, unless otherwise stated, uppercase letters such as P and Q will denote *statements* rather than classes of objects. So P could be

the area of a square having side length L equals L^2

which is a true statement, and Q could be

-4.2 is a positive integer

which is a false statement. A compound statement of the form

If P, then Q.

is a *conditional statement*; the statement P is its *antecedent* and Q is its *consequent*. Take a moment to memorize these terms; they are standard and will occur repeatedly in the next few pages.

Example. Consider the conditional statement

If I am a lawyer, then I can understand legal contracts.

Let's break it down:

If <u>I am a lawyer</u> , then <u>I can understand legal contracts</u> .
  antecedent, P consequent, Q

Note that we are not free to interchange the antecedent and consequent of a conditional statement without a (possibly drastic) change in meaning.

Example. The statement

If I can understand legal contracts, then I am a lawyer.

is clearly false; many engineers can understand legal contracts. This statement is called the *converse* of the statement in the preceding example.

In Chapter 7, we will have more to say about the converse of a conditional statement; we'll also cover two related statements called the *inverse* and *contrapositive*. Of these, only the contrapositive is equivalent to the original conditional. We now examine some argument forms involving conditional statements. Please don't be discouraged by their technical sounding names. Arguments of these forms are recognizable in all engineering discourse.

Modus Ponens, or Affirming the Antecedent

The pattern

If P, then Q.
P.   ✓ valid
Therefore, Q.

is a standard argument form called *modus ponens*. Since the antecedent of the conditional in the first premise is affirmed by the second premise, the form is also called *affirming the antecedent*.

Example. The argument

If our proposal contains errors, we won't receive funding. Our proposal does contain errors. So we won't receive funding.

takes the form of modus ponens. It is a valid argument.

Modus Tollens, or Denying the Consequent

The following pattern is a standard argument form called *modus tollens*. By *not-Q*, we mean the statement called the *negation of Q*.

If P, then Q.
Not-Q.   ✓ valid
Therefore, not-P.

In English we can negate a statement by appending *It is false that* to the start of it, although this may not yield the most concise or graceful formulation.

Example. The statement

All real numbers are positive.

is false. Its negation can be phrased in any of the following ways:

It is false that all real numbers are positive.
Not all real numbers are positive.
Some real numbers are not positive.
There is at least one real number that is not positive.

The negation of a false statement is true, and the negation of a true statement is false. Let's get back to modus tollens. Since the consequent of the conditional in the first premise is denied by the second premise, the form is also called *denying the consequent*.

Example. The argument

If the machine works properly, its output exceeds two units per hour. Its output does not exceed two units per hour. So the machine does not work properly.

takes the form of modus tollens. It is a valid argument.

An Argument Form with Two Conditional Premises

Here's another standard reasoning pattern, this time with conditional statements for both premises.

If P, then Q.
If Q, then R.   ✓ valid
Therefore, if P then R.

Example.

If V_0 exceeds 10 V, the output current will exceed 1 mA. If the output current exceeds 1 mA, the system will fail. We conclude that if V_0 exceeds 10 V, the system will fail.

Fallacies Involving Conditional Statements

What can go wrong with arguments containing conditional statements? There are two famous fallacies to guard against. The first is called *affirming the consequent*.

Example. The argument

If the voltage is high, the current is low. The current is low. Therefore the voltage is high.

is *not* modus ponens. The antecedent of the conditional ("the voltage is high") is not affirmed in the second premise; rather, the consequent ("the current is low") is affirmed. This argument is invalid.

In general, an argument of the following form is fallacious.

If P, then Q.
Q.   × invalid
Therefore, P.

Again, this is called affirming the consequent. Let's proceed to the second fallacious form.

Example. The argument

If our design is the best, we will win the competition. Our design is not the best. Therefore we will not win the competition.

is *not* modus tollens. The consequent of the conditional ("we will win the competition") is not denied; rather, the antecedent ("our design is the best") is denied. This argument is invalid.

In general, an argument of the following form is fallacious.

If P, then Q.

Not-P.   × invalid

Therefore, not-Q.

This is called *denying the antecedent*.

The Disjunctive Syllogism

Another valid argument form, commonly seen, is the *disjunctive syllogism*. The pattern is

P or Q.

Not-P.   ✓ valid

Therefore, Q.

The first premise guarantees that at least one of the statements P and Q must hold. This, taken together with the second premise (that P does not hold), is enough to guarantee that Q holds.

Example. The argument

> Either our measured data are wrong, or our analysis method is wrong. Our measured data are not wrong. Therefore, our analysis method is wrong.

takes the form of a disjunctive syllogism. It is valid.

Informal Fallacies

The fallacies presented above are examples of formal, deductive fallacies. Another part of logic, called *informal logic*, collects and classifies other types of fallacies that commonly occur in human discourse. People have committed these types of errors intentionally or unintentionally since the time of Aristotle (384 BC – 322 BC). Avoid them all.

Ad Hominem

We commit this fallacy when we argue against the person by attacking or discrediting him, or alluding to his possible motives.

Example. Here's an ad hominem argument from a student:

> Dr. Smith says that I have an error in my theory. But he gave me a bad grade in his class and has always had it in for me, so I think he is saying this just to hurt me. I don't think I can trust him, so I think he is wrong and I am right.

A professor could award a bad grade and still correctly identify an error in a theory. That's why this argument is fallacious.

Fallacy of Accident

We commit this fallacy when we try to apply a rule to a case it was not intended to cover.

Example.

> For safety, we store acids in glass containers. Hydrogen fluoride is an acid, so we should store it in a glass jar.

While it's true that acids are commonly stored in glass containers, hydrogen fluoride (or hydrofluoric acid) is used to etch glass. It obviously cannot be kept in a glass jar.

Straw Man Fallacy

We commit this fallacy when we distort someone else's position and then attack the distorted version.

Example.

> Ben, the postdoc in our research group, wants me to repeat the same measurement 20 times. Ben always invents busywork for us so he can impress Dr. Handley. I shouldn't have to repeat something 20 times just to make Ben look like he's doing his job. I won't do it, and I won't include it in the technical report.

In fact, Ben wants 20 repetitions as a good statistical sample. His intention is not to manufacture busywork for anyone.

Appeal to Ignorance

We commit this fallacy when we give up on further thinking and investigation. We might say, for instance, that event A must have caused event B because we cannot imagine any other reason for the occurrence of B.

Example.

> We were unable to show that the system is optimal, hence it is suboptimal.

The system could be optimal even though verification of this was seemingly out of reach.

Hasty Generalization

We commit this fallacy when we conclude something about all members of a group from the characteristics of an insufficient sample.

Example.

> The engineers with whom I interacted during my last job were all terrible writers. My new colleague John is an engineer. John must be a terrible writer.

John may turn out to be a great writer.

Post Hoc Ergo Propter Hoc

We commit this fallacy when we assert that event A must have caused event B because A preceded B in time.

Example.

> We lowered the temperature in the room in the morning, and our measurements were better in the afternoon. So measurements are better when the room is cooler.

The conclusion may be correct, but the argument form is still fallacious. Consider:

> I tied my shoes this morning and my measurements improved in the afternoon. Measurements are better when my shoes are tied!

Cum Hoc Ergo Propter Hoc

We commit this fallacy when we assert that event A must have caused event B because A and B occurred simultaneously.

Example.

> The street lamp went off right when I passed underneath. I must
> be controlling street lamps through some paranormal effect.

In fact, *street light interference* is a classic example of observer bias. Another example:

> The brakes in the car failed at the same time the alternator failed,
> so the alternator failure must have caused the brake failure.

In fact this could just be a coincidence, or both incidents could have the same cause.

Fallacy of Composition

We commit this fallacy when we erroneously attribute a trait possessed by all members of a class to the class itself.

Example.

> I checked the manual and found out that the quick-lock vise on the
> milling machine we want to purchase can only be set using metric
> units. I assume this must be a metric-only milling machine. Let's
> not get it.

Just because the attachment is metric doesn't mean the machine itself can't be set using English units.

Fallacy of Division

We commit this fallacy when we erroneously attribute the traits of a class of objects to each of the separate objects.

Example.

> American engineers produce many inventions each year. John is
> an American engineer. Therefore, John produces many inventions
> each year.

The statement refers to American engineers as a group. This group invents many things each year. From this we can draw no conclusion about John.

Begging the Question

We commit this fallacy when we use the conclusion we're trying to establish as one of our premises.

Example.

> The system is highly productive with minimal waste of energy because it is efficient.

The phrase "highly productive with minimal waste of energy" is a synonym for efficient. The argument being made is "The system is efficient because it is efficient." This is begging the question.

Weak Analogy

We commit this fallacy when we argue based on an alleged similarity between two situations that, in reality, are not that similar.

Example.

> Electric current is like water flowing in a pipe, and the battery is like a pump. When an old rusty pump gets clogged, it has trouble pumping water. Therefore, an old battery loses voltage because it gets clogged with electrons that it can no longer pump through the circuit.

In fact, the analogy between electric current driven through a circuit by a voltage source and water driven through a plumbing system by a pump is often used as a first explanation of electricity for children. It could be appropriate as a refresher for non-engineers as well. For the electrical engineer, however, it is certainly a weak analogy.

False Dichotomy

We commit this fallacy when we base an argument on the premise that either A or B must hold, when in reality a third possibility C could hold.

Example.

> The electric field must be either positive or negative, so it is definitely present and affecting our experiment.

In fact, the value of an electric field can be positive, or negative, *or zero*.

Fallacy of Suppressed Evidence

We commit this fallacy when we omit counterinstances while drawing an inductive conclusion.

Example.

> I couldn't adapt the old code for this new project because the vendor stopped supporting the language ten years ago.

OK, but you failed to mention that you're aware of several other versions of the language sold by competitors that should work perfectly well.

Fallacy of Equivocation

We commit this fallacy when we use a word in two different ways in the same argument.

Example.

> Since we built that 1:2 model of our prototype on such a large scale, we had better use the biggest scale we have to weigh it.

Fallacy of Amphiboly

We commit this fallacy when we argue based on a faulty interpretation of an ambiguous statement.

Example. Here's an example of amphiboly and equivocation together:

> I learned that the transistor dissipates 10 watts with the current experiment.

We don't know if it always dissipates 10 W, and you found out by running the experiment, or if it dissipates 10 W only during the experiment. That's amphiboly. We also don't know if you mean the *existing* experiment, or the experiment in which you measure electric current. That's equivocation. Remember, *current* has two meanings.

Appeal to the Crowd

We commit this fallacy when we argue that statement A must be true because most people believe it's true.

Example.

> Almost everybody uses Java to program these types of applets, so Java must be the best language for these applets.

Fallacy of Opposition

We commit this fallacy when we argue that statement A must be false because our opponent believes it's true.

Example.

> I've been working on this theory of friction for years. If you don't agree with it, you've obviously not considered it thoroughly.

Appeal to Authority

We commit this fallacy when we argue that statement A must be true because experts believe it's true.

Example.

> My advisor said that purple fribble has a small Young's modulus, so I could use it to build my device. Then I asked Dr. Smith, who said the Young's modulus isn't as small as my adviser claimed. Unlike my advisor, Dr. Smith is a full professor and department chair. He must be correct, so I'll use something else.

4.3 Additional Checks on Correctness

We have seen that an ability to recognize fallacious argument forms is essential to the engineer. In this section, we touch on other ways of catching errors in claims. We like to class these under the general heading of *physical reasoning*, although (1) many are quantitative or semi-quantitative in nature, and (2) we do not address the special reasoning modes employed by modern physicists. A fancier name is *heuristic reasoning*.

Intuitive Plausibility

Is the answer reasonable?

> **Example.** Suppose we take an isolated particle and apply an external, unbalanced force directed to the right. If our calculation indicates that the particle responds by accelerating to the left, perhaps we should look for an error.

Dimensional Checks

Physical dimensions must appear consistently in a valid physical equation.

> **Example.** Suppose we write
>
> The position along the x-axis of the particle at time t is given by
>
> $$x = x_0 + v_0 t + \tfrac{1}{2}at^2 \qquad (1)$$
>
> where x_0 is the initial position, v_0 is the initial speed, and a is the acceleration (assumed constant).

The physical dimensions that appear in equation (1) are as follows:

$$[\text{length}] = [\text{length}] + \frac{[\text{length}]}{[\text{time}]} \cdot [\text{time}] + \frac{[\text{length}]}{[\text{time}]^2} \cdot [\text{time}]^2.$$

So all terms have the same dimensions (that of [length]), and this is *necessary* for the equation to be correct. Of course, it is not *sufficient*; it does not guarantee, for example, that (a) the numerical coefficient $1/2$ is correct, or (b) there isn't a term missing from equation (1). Nonetheless, routine dimensional checks are a strongly recommended practice.

> **Example.** If w, x, y, z all have units of length, then something is wrong with the equation
>
> $$w = \frac{x^3 - y}{x^2 + z}.$$

Order-of-Magnitude Checks

Quantitative claims should be numerically reasonable.

Example. The electric currents that flow, under normal conditions, in a circuit operated by a small battery are probably in the mA (milliampere) range. They could be an order of magnitude larger or smaller than that, of course. Suppose Bill (an electrical engineering student) calculates a current in his hand-held circuit as follows:

$$I = V/R = 2.5/(10000) = 0.25 \times 10^3 = 250 \text{ A} .$$

Although Bill made a calculation error, he still has a chance to ask whether this answer is physically reasonable. Hint: The current in a cloud-to-ground lightning bolt might peak at several amperes.

Expected Variation with a Parameter of the Problem

Some engineering students associate the term *variable* with the term *letter*. But certain letters are conventionally used to represent true constants: examples are the Greek letter π and the base e of the natural logarithm. We would not call the corresponding quantities "variables" just because they are denoted by letters.

Other quantities in a problem may be denoted by letters but *temporarily held fixed during a calculation involving other variables*. We refer to these quantities as *parameters*.

Example. The equation $y = ax^2$ describes a *family* of parabolas: one for each value of a. To plot one curve from the family, we might set $a = 2$ and plot $y = 2x^2$. To plot another curve, we could set $a = 3$ and plot $y = 3x^2$. In plotting each curve of the family, we are holding a constant. On the other hand, a is not a "true" constant like e or π. It isn't appropriate to call a a variable or a constant: it is an example of a parameter.

Many problems of engineering interest involve parameters. By obtaining an answer in terms of parameters relevant to a problem, and by understanding how the answer to a problem *should* vary with each parameter, we gain an avenue for checking our final answer.

Example. Suppose we wish to find the volume V of a right circular cylinder having radius a and height h. In calculus we learn to set up the integral

$$V = \int_0^h \int_0^{2\pi} \int_0^a r \, dr \, d\theta \, dz$$

for this purpose. Although various letters appear on the right-hand side, they are not all "variables" at this stage of the game. The variables of integration are r, θ, and z; these are actually changing (over the ranges indicated by the integration limits) during the integration process. But h and a are fixed during the integration and are properly regarded as parameters. Completing the integration, we get an answer in terms of these parameters:

$$V = \pi a^2 h \,.$$

Now we can choose to mentally vary each parameter while holding the other one fixed. Holding a fixed and increasing h, we find that V increases according to this formula. This is, of course, as expected for a cylinder that is getting taller. Similarly, if we hold h fixed and decrease a, we find that V decreases as expected. Engineers should always be thinking in this way: we never tell our students they should reach an answer, box it in, and move on without mentally playing with each parameter in the answer. If they followed our advice, they would never be satisfied with answers such as

$$V = \pi a^2 / h$$

to the problem at hand.

The above example shows the value of solving problems using parameters. We could have treated a cylinder 1 m high and 10 cm in radius, thereby considering just one specific case. Instead, by using parameters, we obtained a formula giving the volume of any right circular cylinder.

Agreement with Known Special Cases

One advantage of working problems in terms of parameters is that a problem may have limiting cases whose answers are known.

Example. Suppose you must find the radial component of the electric field in the bisecting plane of a uniformly charged line segment of length $2L$. After a long calculation, you arrive at

$$E_\rho(\rho) = \frac{\lambda}{2\pi\epsilon_0\rho} \cdot \frac{\rho^2 + L}{(\rho^2 + L^2)^{1/2}}$$

where λ is the charge density, ρ is the perpendicular distance from the field point to the segment, and ϵ_0 is a constant called the free-space permittivity. Hoping for a quick quality check on this answer, you consider the trivial case in which $L = 0$; this will make the charged segment disappear and should yield a null result for the electric field. Unfortunately, you obtain

$$\lim_{L \to 0} E_\rho(\rho) = \frac{\lambda}{2\pi\epsilon_0 \rho} \cdot \frac{\rho^2}{(\rho^2)^{1/2}} = \frac{\lambda}{2\pi\epsilon_0} \neq 0 \,.$$

There must be a calculation error. After finding and correcting the error, you arrive at

$$E_\rho(\rho) = \frac{\lambda}{2\pi\epsilon_0 \rho} \cdot \frac{L}{(\rho^2 + L^2)^{1/2}} \,,$$

which does behave correctly as $L \to 0$. Seeking a further check, however, you once again hold ρ fixed and let $L \to \infty$ to get

$$E_\rho(\rho) = \frac{\lambda}{2\pi\epsilon_0 \rho} \,. \tag{2}$$

Referring to an electromagnetics handbook, you find that this is indeed the field at distance ρ from a uniform line charge of infinite length. This is good news, but we still caution that it isn't a guarantee as many similar and not-so-similar expressions also reduce to (2) as $L \to \infty$. However, if your final answer *failed* to reduce to the known result (2), you'd know that a mistake was made somewhere (a mistake by you, by the person who derived (2), or both). Checking for agreement with known special cases is just one more tool you can use to hunt for errors in claims before you write those claims in your engineering document.

You can also run checks by approximating an answer: dropping small terms, ignoring slow time variations, etc. Such techniques often receive extensive coverage in engineering courses.

Other Mathematical Properties of the Answer

If an answer is time dependent, you might check its initial value, final value, or time-average value for correctness.

Example. Suppose your answer to a problem is a time function

$$y(t) = 4 + \cos 200\, t \,.$$

The average value of $y(t)$ is 4. If this seems wrong, you have reason to look for a calculation error.

Another good check is to look for inappropriate singularities.

Example. When solving the wave equation in cylindrical coordinates, it is often inappropriate to keep the Hankel function solution since it has a singularity on the z-axis. You may remember that this situation arises when finding the field inside a circular waveguide. If you cannot provide a physical reason why an electric field should be infinite, then your answer could be wrong.

Accord with Standard Physical Principles

Certain notions, such as causality and symmetry, are encountered routinely in physics. Why not use them to check your answers whenever possible?

Example. Suppose you must find the magnitude of the electric field at distance r from a point charge Q. After a page of calculations, you arrive at

$$E(r, \theta) = \frac{kQ}{r^2 + \theta^2}$$

where k is a constant and θ is the polar angle of spherical coordinates. Well, aside from the obvious problem with dimensions (you cannot add a distance squared to an angle squared), there is a problem with symmetry here. If we hold r fixed and vary θ, we are physically walking around the point charge while staying at constant distance from it. The symmetry of this extremely simple charge distribution implies a symmetry in the resulting electric field: it should *not* vary with θ. This gives you another reason to scrutinize your calculations.

Example. Suppose you seek the response of a physical system to a given input, and by a long calculation find that the response begins *before* the input is applied. This violates the accepted physical principle of *causality*: effects cannot precede their causes. Better look for a calculation error.

Another important principle is superposition, although one must remember that it applies only to linear systems.

Example. Often the whole cannot be greater than the sum of its parts. If you use 1000 tons of steel and 500 tons of concrete to construct a bridge, it is doubtful if the resulting bridge weighs 5000 tons (unless you have neglected to consider a component).

In contrast, the whole is often *smaller* than the sum of its parts because of cancellation. Two very large forces acting on a single bolt may produce little torque if the moment arms are the same and the forces are applied in opposite directions.

4.4 Other Ways to Be Careful

There are many ways to actively guard against the kind of sloppiness that leads one to make errant claims. Here are some suggestions.

1. Do not jump to conclusions. Stop and think. Review all available information. Seek expert help if necessary. Cover all bases before making a claim.

2. Maintain a critical attitude. Guard against distortions, whether honest or dishonest. Maintain reasonable skepticism even about the literature of your own field.

3. Respect the truth. You are not trying to find evidence to prove one of your preconceived notions. Rather, use an unbiased, calm mind to listen to what the evidence actually says.

4. Remember Ockham's razor. Prefer the simplest design or explanation.

5. Insist on reliable evidence from dependable sources. Your cubicle-mate's belief in something may not suffice for your purposes (that is, for your reader's purposes). He or she believes that electromagnetic waves travel faster than the speed of light? Perhaps this should be verified by an expert authority.

6. Double check everything! This takes additional time and effort, of course, but it may save the reader from having to evaluate false claims.

7. Always look for counterexamples to your claims. Look at each claim with a critical eye, trying to construct a counterexample. Any astute reader will be doing the same thing when reading your document.

> **Example.** The *post hoc ergo propter hoc* and *cum hoc ergo proper hoc* fallacies should have you on alert about jumping to conclusions regarding cause-and-effect relations. Remember the old saying:
>
> > *Correlation does not imply causation.*
>
> Sure, we may notice a strong positive correlation between two events X and Y. It *could* be that X causes Y. But it could also be the case that Y causes X, that both X and Y are among the effects of some cause Z, or that the observed correlation is just accidental. It often takes careful, planned experimentation to sort out cause-and-effect relations. An engineer's gut intuition cannot always be trusted in such matters.

4.5 Chapter Recap

1. Consideration of standard fallacies can teach us a lot about common thinking blunders.

2. Formal logical fallacies include denying the antecedent and affirming the consequent.

3. Informal fallacies include things like *ad hominem*, straw man, and appeal to ignorance. A quick summary of these fallacies appears on p. 155.

4. Many techniques are available for checking claims before the reader sees them.

5. The engineer should maintain a critical attitude and a respect for truth. One way to be critical of a claim is to seek counterexamples.

6. Don't jump to conclusions. In particular, correlation is not causation.

7. Ockham's razor (also spelled *Occam's razor*), a long-time favorite in scientific reasoning, is the principle that explanations should not be more complicated than necessary.

8. Spending the time to double check every claim is better than misleading the reader and becoming embarrassed in print.

4.6 Exercises

4.1. Identify the argument as valid or fallacious.

(a) All waveguides are inefficient. This resistor is inefficient. Therefore, this resistor is a waveguide.

(b) All ceramic capacitors are non-polarized. This capacitor is non-polarized. This capacitor must be a ceramic capacitor.

(c) All gold is diamagnetic. This metal is diamagnetic. It must be gold.

(d) All transistors are made of semiconductors. A diode is not a transistor. Therefore, a diode is not made of semiconductors.

(e) All rich people are happy. Some engineers are rich. Therefore, all engineers are happy.

(f) All Fourier transformable functions have a finite number of discontinuities in a given interval. The function $f(x) = x^2$ has a finite number of discontinuities in a given interval. Therefore, $f(x) = x^2$ is Fourier transformable.

(g) All passive two-port networks have $|S_{21}| \leq 1$. My network has $|S_{21}| \leq 1$. Therefore, my network is passive.

(h) All resistors are marked with color codes. This component has a color code. Therefore, this component is a resistor.

(i) All 3.5 mm connectors are precision connectors. Some RF connectors are 3.5 mm connectors. Therefore, all RF connectors are precision connectors.

(j) All K-connectors are precision connectors. All K-connectors are mechanically compatible with 3.5 mm connectors. Therefore, all connectors that are mechanically compatible with 3.5 mm connectors are precision connectors.

4.2. State whether the following argument forms are valid. Assume P, Q, R, S are statements.

(a) P or Q.
        If P, then R.
        If Q, then R.
        Therefore, R.

(b) P or Q.
        If P, then R.
        If Q, then S.
        Therefore, R or S.

(c)   Not-R or not-S.
        If P, then R.
        If Q, then S
        Therefore, not-P or not-Q.

4.3. Consider the list of *invalid* categorical syllogisms shown on p. 154. For each syllogism, find a real-world counterexample that shows the syllogism is invalid.

4.4. Look for fallacies.

(a) Smith's results are questionable because he has made significant errors in the past.

(b) Since it is impossible to conceive of anything but electromagnetic interference causing this problem, the problem must be due to electromagnetic interference.

(c) Having discovered failures in two of the modules tested at random, we concluded that all 10,000 modules likely failed.

(d) The system temperature increased after we heard the noise, hence the noise must have caused the temperature increase.

(e) Transistor leads are like tiny legs. Since people have two legs, transistors have two leads.

(f) Only two possibilities exist: either the temperature decreased or it increased. Since both of these represent changes, we do know that the temperature changed over time.

(g) This transistor is superior to the other alternatives because it is better.

(h) Resistors often have green stripes. Therefore, they seldom have blue stripes.

(i) All machines are somewhat inefficient. Sam is somewhat inefficient. Therefore, Sam is a machine.

(j) A resistor is an electrical device. A transistor is an electrical device. Therefore, a resistor is a transistor.

(k) There must be something wrong with subsystem A. Ever since it was redesigned, subsystem C has been unreliable.

(l) This new system is unreliable. Out of the 10000 units delivered to us, two random units were chosen for testing and both failed.

(m) We connected a 10 V capacitor and it exploded! We must have exceeded the voltage rating.

4.5. As humans, we have *cognitive biases* that lead us to distort our experiences and process information selectively. Do some background reading about cognitive biases. Could any of these patterns make it easier to commit fallacies?

4.6. Consult a logic textbook to learn the Venn diagram method for validating syllogisms. Use the method to validate the 19 syllogisms listed on p. 153.

5

Some Points of Grammar and Style

We have likened the rules of writing to engineering standards. In this chapter we review some basic aspects of English style necessary for technical writing. Rather than a formal exposition, we offer a list of suggestions much in the form of the classic book *The Elements of Style*, written by William Strunk Jr. nearly one hundred years ago. We try to minimize grammatical terminology and leave a review of essential terms for the Quick Reference (see p. 167).

5.1 Rules and Suggestions

Here's our short list of rules and suggestions regarding grammar and style. Our experience is that even advanced students have difficulty with these topics. Upon mastering these you'll be on good footing, and can proceed to some of the more subtle suggestions of William Strunk and others.

The paragraph is the unit of exposition.

The sentences in a paragraph should all hang together somehow.

> **Example.** Let's look at the first two paragraphs of Chapter 2 of the book *Electromagnetics* (2nd edition, Taylor & Francis CRC Press, Boca Raton, FL, 2008) by the authors:
>
> > In 1864, James Clerk Maxwell proposed one of the most successful theories in the history of science. In a famous memoir to the Royal Society he presented nine equations summarizing all known laws on electricity and magnetism. This was more than a mere cataloging of the laws of nature. By postulating the need for an additional term to make the set of equations self-consistent, Maxwell was able to put forth what is still considered a complete theory of macroscopic electromagnetism. The beauty of Maxwell's equations led Boltzmann to ask, "Was it a god who wrote these lines ... ?"

The second paragraph starts as follows. We have used the first paragraph to introduce Maxwell's theory, and now switch to thinking about how to best present it.

> Since that time authors have struggled to find the best way to present Maxwell's theory. Although it is possible to study electromagnetics from an "empirical–inductive" viewpoint (roughly following the historical order of development beginning with static fields), it is only by postulating the complete theory that we can do justice to Maxwell's vision. . . .

We hope you have an intuitive sense of what a paragraph should be. Every paragraph needs a purpose, and each sentence it contains should contribute to the fulfillment of that purpose. Don't sprinkle indents randomly through a document. Some books on technical writing will argue that an indent here or there is ok just to give the reader a rest. On the contrary, we urge you to study your material until you discover a way to decompose it meaningfully and naturally into reasonably-sized true paragraphs.

Don't write sentence fragments. A declarative sentence needs a subject and a predicate.

A *declarative sentence* makes a statement (rather than, say, asking a question, giving a command, or making an exclamation). Most of the sentences in a formal engineering document should provide information and therefore should make statements. A declarative sentence needs a subject and a predicate. The *predicate* spells out the action and contains the verb, while the *subject* is the person or thing doing the action.

> **Example.** You should be able to identify the subject and predicate portions of any declarative sentence you write. In a sentence arranged naturally, the subject comes first.
>
> A keystroke monitor runs continuously in the background .
>
>   subject   predicate
>
> A flush-mount antenna is needed for this particular application .
>
>   subject   predicate

Example. Inverted sentence arrangements are possible but are best used with care. The following is acceptable:

What a difficult design task this was.

It should be clear which of the following sentences is best:

✓ This analysis was complicated.

Complicated was this analysis.

Dialogue of the latter type was used in the *Star Wars* movies to make the character Yoda sound exotic. Technical writing should not sound exotic.

Example. Here are examples of *sentence fragments*:

The total force $F = kx$ acting on the spring.

Where k is the spring constant.

They are punctuated as sentences, but each lacks the needed subject-predicate combination.

In addition to declarative sentences, a document may contain *imperatives*. An imperative sentence expresses desire, permission, or a command.

Example. The following statements are imperatives:

Note that $f = 2$ unless $p = 1$.

Observe that $k = 0$ in this case.

Suppose $h = 4$.

Set $q = 2$.

Choose any positive integer n.

In these cases the implied subject is *you*.

Avoid excessive sentence length.

Extremely long sentences can be hard to read.

Example.

The Hertz potential Green's dyad, an extremely useful tool in the analysis of high-speed integrated electronic circuits incorporating multilayered and possibly complex electromagnetic media, can be conveniently, and yet rigorously, expressed in terms of an integral representation involving the ordinary Riemann integral as defined in calculus.

✓ The Hertz potential Green's dyad is an extremely useful tool in the analysis of high-speed integrated electronic circuits. Such circuits may incorporate multilayered and possibly complex electromagnetic media. The dyad can be conveniently, and yet rigorously, expressed in terms of an integral representation; this representation involves the ordinary Riemann integral as defined in calculus.

Although short sentences can be easy to read, a succession of these can make it sound like you're writing for elementary school students.

Example. Consider

We turned the system on. The green light flashed. We measured the output voltage. It was 5 V. Then we measured the current. It was 4 mA. Ohm's law held.

Despite the fact that *brevity is the soul of wit* (Shakespeare's *Hamlet*), it's unwise to sound like a third grader.

Don't write run-on sentences.

A run-on sentence will take one of two forms. In the first form, independent thoughts are smeared together without punctuation.

Example. Here is a run-on sentence:

The design was easy it took only a few days to complete.

Here's how it could be fixed:

The design was easy; it took only a few days to complete.

We'll say some things about semicolons later.

In the second form, independent thoughts are spliced together with a comma but without a conjunction.

Example. This is also a run-on sentence:

The design was easy, it took only a few days to implement.

Here's how it could be fixed:

The design was easy, and it took only a few days to implement.

Make the important ideas prominent grammatically.

By failing to consider the order of words in a sentence, we can force the reader to work harder than necessary.

Example. Consider the sentence

The main component of steel is iron.

The arrangement here is natural, with the subject (*The main component of steel*) coming first. But there are really three ideas here: *component*, *steel*, and *iron*. Of these, the two most important are *steel* and *iron*; the reader shouldn't have to wait until halfway through the sentence to reach them. Compare with

✓ Steel is mostly iron.

In this case the idea of *main component* (embodied here as *mostly*) has been properly subordinated to the two important ideas.

Example. By starting a sentence with *It is*, *It was*, *There is*, *There are*, *There were*, or *There have been*, you force the reader to wait for your main ideas.

It is noted that for details of the analysis, the reader may consult Lu [12].

✓ Details of the analysis appear in Lu [12].

There is another parameter called resistance, defined as $R = V/I$.

✓ A parameter called resistance is defined as $R = V/I$.

There are three components of a radio link: a transmitter, a channel, and a receiver.

✓ A radio link has three components: transmitter, channel, and receiver.

Use (but don't overuse) the active voice.

Voice is a form of a verb indicating the relation of the subject to the action performed. In the *active voice*, the subject takes action. In the *passive voice*, the subject receives the action.

Example. Active voice:

> We thoroughly analyzed the system.

We (subject of the verb) did the analysis. The *system* (object of the verb) got analyzed. Passive voice:

> The system was thoroughly analyzed by us.

The system (subject of the verb) got analyzed (by us). Notice how lifeless this sounds.

Example.

> The system was turned on by us. Then, the output was measured by a voltmeter.

> ✓ We turned the system on and measured the output with a volt-meter.

Like everything else in writing, however, the active voice shouldn't be overused. *Sentence variety* is an important aspect of good classical composition. See p. 158 for a list of 18 ways to start a sentence.

Consider the subjunctive mode for conditions contrary to fact.

Although the grammatical term *subjunctive mode* may sound unfamiliar, you'll recognize the mode when you see it.

Example. Suppose we're trying to argue by contradiction. Maybe our argument starts this way.

> The number $\sqrt{2}$ is not a rational number. For if it were, it could be written as m/n for some integers m and n ...

The phrase *if it were, it could* puts us in the subjunctive mode. This is appropriate for asserting something like $\sqrt{2}$ *is rational*, which is contrary to fact (and will eventually lead to a contradiction; see the discussion of this on p. 134). The following version should sound much less appropriate:

> The number $\sqrt{2}$ is not a rational number. For if it is, it can be written as m/n for some integers m and n ...

How can we say *if it is* when we just said that *it isn't*? Proof by contradiction will be discussed further in Chapter 7.

The subjunctive also appears in such phrases as

> Suffice it to say, ...

Nonetheless, the subjunctive has fallen out of favor in recent years and many published writers avoid it. We agree that people probably shouldn't be writing

> For if P be any point on the line segment ...

anymore (there is no doubt that someone could find a point on a given line segment and call it P). But for conditions contrary to fact, the subjunctive can sound more logical.

> **Example.** Electrolysis of water is not an efficient means for producing hydrogen. If it were, it would be widely used in industry as it is environmentally friendly.

Punctuate correctly.

Entire books have been written on punctuation, and many English dictionaries contain useful punctuation guides. Here's a quick survival guide:

1. End each declarative sentence with a period.

2. End each question with a question mark.

3. The comma produces a short pause. Use commas to separate words, phrases, or clauses in cases where confusion could result otherwise.

4. Use a colon to introduce a list or an explanation of what precedes it.

5. Use a semicolon to separate complete but related independent thoughts.

6. Use a dash to introduce a break in a sentence.

7. Use an apostrophe to show possession.[1]

---

[1]Or to form a contraction such as *can't*, but we suggest you use contractions sparingly in formal technical documents. We also implore you to control any urge to use exclamation points — except in those rare cases where emphasis is imperative!

8. Use a hyphen to break a word at the end of a line, or to conjoin certain words.

9. Use an ellipsis to indicate that you have excluded material from a quoted passage.

Example. These declarative sentences are all punctuated properly:

> Specific examples of this procedure will be given in Chapter 4.
>
> The main ideas were due to Wallis, Newton, and Leibniz.
>
> The required construction is, however, lengthy and tedious.
>
> We now address our main problem: how to get the engine running.
>
> There are three main issues: efficiency, effectiveness, and cost.
>
> We have found a solution; indeed, direct substitution verifies this.
>
> The manufacturer's instructions prohibit such usage.

Example. You should end a long introductory phrase with a comma.

> As a condition for effective combustion and adequate efficiency the level of oxygen must be appropriate.
>
> ✓ As a condition for effective combustion and adequate efficiency, the level of oxygen must be appropriate.

A short introductory phrase may not require a comma. The sentence

> Upon seeing the result we understood our error.

is fine the way it is, although one could insert a comma between *result* and *we* if desired.

Example. You should nest a parenthetical expression between commas. This sentence contains a parenthetical expression:

> It is the electric current, which is the rate of flow of electrons through the filament, that determines the light intensity emitted by the bulb.

The expression between commas is *parenthetical* because it could be deleted without changing the essential nature of the sentence (it is extra information).

Don't create compound words by impulsively connecting various parts of speech with hyphens. The rules for hyphenation in English are too complex to fully discuss here; when in doubt, consult a dictionary or style guide. We'll simply mention a couple of all-too-common errors.

Example. You should *not* unite a noun with its adjective modifier using a hyphen:

We display the performance of the receiver front-end in real-time.

Compare:

Let $\Re(z)$ denote the real-part of a complex number z.

✓ Let $\Re(z)$ denote the real part of a complex number z.

You can hyphenate adjectives modifying a noun if they come before the noun, as in

✓ high-frequency oscillations

✓ water-based solvent

but not if they come after the noun, as in

✓ a solvent that is water based.

Punctuation often determines the meaning assigned to a sentence. Anything you choose to learn about proper punctuation will pay dividends in the long run.

Use correct capitalization.

You should capitalize the first word in a sentence, a proper name, a book title, an academic title, a day of the week, and so on, according to accepted convention. Don't capitalize a word simply because it's important.

Example. Do these things:

This result is equivalent to (1.7).

The Bessel function $J_0(x)$ must be employed for this purpose.

See *Advanced Calculus* by R.F. Smith.

We expect to have a final result by the end of June.

Don't do these things:

This effect was due to Interference from an electromagnetic wave.

We consulted an expert from Academia.

This approach represents Standard Practice in Industry.

We measured the thickness using a Digital Micrometer.

Proper capitalization of unit symbols is important. The first letter of a unit symbol is capitalized only if the unit is a proper name. The symbol T denotes Tesla, where the symbol t denotes tonne (1000 kg). Improper capitalization of prefixes can lead to embarrassing errors — imagine misleading the reader with the symbol MJ (*mega*joules) when you meant mJ (*milli*joules). Standard unit symbols and prefixes are listed on p. 160 and p. 161.

Be concise.

Don't saddle sentences with excess baggage. As William Strunk says, "Omit needless words."

Example.

The machine has proved itself to be efficient.

✓ The machine has proved efficient.

A short list of fluff phrases — such as *the reason is because* and *as a matter of fact* — appears on p. 161.

Example. Think twice before leaving the words *reason* and *because* in the same sentence.

The reason the system failed is because part *A* failed.

✓ The reason the system failed is that part *A* failed.

✓ The system failed because part *A* failed.

Example. State facts rather than labeling statements as facts.

It is a fact that electric charge is a basic property of matter.

✓ Electric charge is a basic property of matter.

Example. Here are some redundant expressions. The strikethroughs show which words can be omitted.

> We found inefficiencies throughout the ~~entire~~ system.

> The singularity is ~~known to be~~ removable.

See the Quick Reference for more examples.

Be direct.

As Strunk says, "Put statements in positive form."

Example.

> To avoid the possibility of not having enough fuel, ...

> ✓ To make sure we had enough fuel, ...

Use parallel construction.

Nontechnical but classic examples of what grammarians call parallel construction include

> I came, I saw, I conquered.

> Give me liberty or give me death.

The first is traditionally attributed to Julius Caesar, the second to Patrick Henry. Imagine, instead,

> I came, saw some things, then got around to conquering.

> Give liberty to me or provide me with death.

Hear the differences? The original versions are more forceful, more memorable, more readable. Don't underestimate the power of parallel construction in English.

Example.

> ✓ Formerly these devices were made from mylar; now they are made from teflon.

> Formerly these devices were made from mylar; in the present day, however, teflon is used to manufacture them.

Use standard abbreviations correctly.

Study the following table and keep it handy.

abbreviation	meaning
etc.	and other things
viz.	namely
e.g.	for example
i.e.	that is
cf.	compare
n.b.	note well
et al.	and others
pp.	pages
ff.	and the pages following

Example. Note that *etc.* means *and other things*. Would it be appropriate here?

We can count like this: one, two, three, etc.

Answer: no. That writer should have finished her sentence with *and so forth*. What *other things*? Spaceships? Apples?

Example. Do not use *for example* and *etc.* in the same sentence. Does this sound right?

The types of capacitors are, for example, mylar, ceramic, tantalum, paper, etc.

Note that *cf.* does not mean *see*. Similarly, *e.g.* does not mean *that is*, and *i.e.* does not mean *for example*.

Example. Avoid the strangely popular construct *i.e., e.g.,* as in

There are many types of screw drives, i.e., e.g., Philips, slot, square and Torq-set.

It's just wrong.

Avoid the pattern *of the ... of the ... of the ...*

This pattern can arise from the translation of heavily inflected languages (such as Russian) into English. The English version can be hard to read.

Example. Consider

> An investigation of the technique used for the computation of f revealed the importance of the configuration of the set of holes relative to the number of failures experienced over the lifetime of the device.

There must be a better way to say this. Sometimes we can revise out *of the* altogether. Compare:

> The integrity of the gate depends on the method of installation.
>
> ✓ The gate integrity depends on the installation method.

Make proper use of comparatives.

Comparatives are words like *bigger, smaller, higher, lower, faster, slower, brighter, darker, heavier,* and *lighter.* Don't use them unless the standard of comparison is stated or understood.

Example. Consider

> We needed to choose a good starting point for our design. The X45 module provides for greater stability.

Greater than what?

Implement grammatical agreement in number.

English words can take different forms to indicate whether one or more entities are under discussion. A verb, for example, must agree in number with its subject.

Example.

> In this chapter, the experimental procedure and setup for the three cases is described.
>
> ✓ In this chapter, the experimental procedures and setups for the three cases are described.

Similarly, a pronoun must agree in number with its referent.

Example.

Each engineer on the team did their best.

✓ Each engineer on the team did his or her best.

The plural pronoun *their* does not match the singular *engineer*.

Remember, the Latin derived words *data, criteria, media,* and *curricula* are plural. The singulars are *datum, criterion, medium,* and *curriculum,* respectively.

Example. The following are incorrect:

The data is shown in Table 1.

What is the most important single criteria?

I always prefer a written media for communicating my ideas.

Typically the plural forms are misused.

Use clear pronoun referents.

A pronoun takes the place of a noun; i.e., it must refer back to some noun introduced previously. If the pronoun and its noun are too widely separated in your text, or if you are careless in some other way, the reader may get confused.

Example. In the second sentence of the following passage, the pronoun *It* clearly refers back to the noun *linkage A*.

The failure was caused by linkage A. It was simply a weak structural member.

Now consider

Therefore we have both $P = I^2 R$ and $W = PT$. Here we have used the definition of power, the meaning of resistance, and the relation between power and energy. Fortunately we are almost done with the derivation. This equation shows that ...

This equation? Which equation? What equation?

Don't misplace modifiers.

Adjectives modify nouns and pronouns. *Adverbs* modify verbs, adjectives, and other adverbs. You should try to place a modifier close to the thing it modifies.

Example. Consider

Just a diode was added to improve circuit safety.

A diode was just added to improve circuit safety.

Note that we changed the meaning of a sentence merely by moving the adverb *just.*

We increased the level of disturbance that we had applied slowly.

We slowly increased the level of disturbance that we had applied.

The first one is probably not what the author means. What would it mean to apply a disturbance slowly?

Watch your verb tenses.

Tense is the property of a verb that indicates *when* (in time) the relation exists or the action is performed. Don't shift tense around impulsively.

Example.

The accurate measurement of force required the use of techniques that satisfy published standards.

✓ The accurate measurement of force requires the use of techniques that satisfy published standards.

The first writer shifts from the past tense *required* to the present tense *satisfy.* Such unnecessary shifts should be avoided.

Some writers have trouble deciding whether to write in present or past tense, so their readers get yanked around temporally.

Example.

The antenna is pointed at the target, and the *S*-parameters are measured. Twenty data sets were taken. Then the FFT is computed.

✓ The antenna is pointed at the target, and the *S*-parameters are measured. Twenty data sets are taken. Then the FFT is computed.

✓ The antenna was pointed at the target, and the *S*-parameters were measured. Twenty data sets were taken. Then the FFT was computed.

Note that the general truth requires the present tense.

✓ An expert informed us yesterday that electric currents produce magnetic fields.
✓ The report I wrote last year is 37 pages long.

Avoid dangling constructions.

A *participle* is a verb form used as an adjective. When a participle starts a sentence, it should be followed immediately by some indication of the *agent* (the entity that performed the action indicated).

Example.

Having substituted (4) into (5), the result is (6).
✓ Substitution of (4) into (5) yields (6).
✓ Having substituted (4) into (5), we obtain (6).

In the first version, the participle is *dangling*; the agent (he or she who did the substituting) is not indicated. The second version does not begin with a participle. In the last version, *we* are the agent so the participle is not dangling.

A *gerund* is a verb form used as a noun. When a gerund occurs at the start of a sentence, it should be followed immediately by some indication of the agent!

Example.

By adjusting the aperture, more light can fall on the sensor.
✓ By adjusting the aperture, one can permit more light to fall on the sensor.

An *infinitive* is the *to*-form of a verb. Don't dangle infinitives.

Example.

> To design the circuit, SPICE must be available.
>
> ✓ To design the circuit, we must have SPICE available.

SPICE (a circuit analysis computer program) is not going to design the circuit. Rather, *we* are going to do so (with the help of SPICE).

Use *due to* as a predicate adjective.

Don't start a sentence with *Due to*. A predicate adjective belongs in the predicate where it can modify the subject of the sentence.

Example.

> Due to this failure, the machine had to be replaced.
>
> ✓ Owing to this failure, the machine had to be replaced.
>
> ✓ Because of this failure, the machine had to be replaced.
>
> ✓ The success of our project was due to the collaborate technology we used.

In the last version, the main verb *was* signals the start of the predicate. This is where *due to* belongs.

Be careful with *neither/nor*.

The pair *neither/nor* requires the singular.

Example.

> Neither m nor n are even integers.
>
> ✓ Neither m nor n is an even integer.

Moreover, *neither/nor* must connect elements of the same grammatical type.

Example.

> This design is neither efficient nor functions properly.
>
> ✓ This design is neither efficient nor functional.

In the first version, an attempt has been made to correlate the adjective *efficient* with the verb phrase *functions properly*. In the second version, *efficient* is correlated with another adjective (*functional*).

The same principle holds for the other correlative conjunction pairs *either/or*, *both/and*, and *whether/or*.

Use *different from*, not *different than*.

You wouldn't say *A differs than B*. You'd say *A differs from B*. So don't say *A is different than B*. Say *A is different from from B*.

Example.

The field inside the cavity is different than the field outside.

✓ The field inside the cavity is different from the field outside.

✓ The field inside the cavity differs from the field outside.

One person can, of course, *differ with* another. *Different* is not the only English word that requires a particular preposition when used with a preposition. Some other pairings are listed on p. 162, but consult a dictionary when in doubt.

Example. Consider the word *speak*. We speak *to* an audience. We speak *to* or *with* a person. We speak *on* or *about* a topic. We speak *at* an event. The words *to*, *on*, *with*, *about*, and *at* are prepositions, but they are not interchangeable in this case.

Use *respectively* to properly correlate items in ordered sequences.

Proper association of items within a list to characteristics in a second list may be accomplished using the word *respectively*.

Example.

✓ The cinch bolt, the latching bolt, and the safety bolt must be torqued to 100, 150, and 275 ft-lb, respectively.

The cinch bolt, the latching bolt, and the safety bolt must be torqued to 100, 150, and 275 ft-lb.

The first version specifies that the cinch bolt must be torqued to 100 ft-lb, the latching bolt must be torqued to 150 ft-lb, and the safety bolt must be torqued to 275 ft-lb. With the second version, the reader might think that, for example, the latching bolt must be torqued to 100 ft-lb. Obviously, correct torque is crucial for proper operation of a fastener.

Avoid double indications of past time.

Often it is necessary to speak about the past while referring to something in an even earlier past.

Example.

We would have liked to have designed a system at lower cost.

✓ We would have liked to design a system at lower cost.

✓ We would like to have designed a system at lower cost.

The first example contains what we are calling a "double indication of past time."

Follow the standard rules for expressing numbers.

Spell out any whole number less than 10 and any whole number that begins a sentence. Otherwise use numerical figures.

Example.

✓ We studied five oscillation modes of the structure.

✓ Fifteen auxiliary pumps failed that same day.

✓ Of these, 78 units were defective.

However, it often makes more sense to spell out a number if it is meant as a rough estimate.

Example.

✓ Electrobop theory has been around for three-hundred years.

✓ Twenty years from now all units will have this safety feature.

There are other exceptions, as for the names of figures.

Example.

✓ See Figure 3.

Format enumerations and lists correctly.

The numbers or letters used to enumerate or list items should be enclosed in parentheses. If the items themselves are punctuated by commas, then semi-colons should be used as separators; otherwise, commas can be used as separators.

Example.

✓ Find (a) the electric field, assuming a uniform charge density; (b) the magnetic field, assuming a uniform current; and (c) the Poynting vector, assuming a lossless medium.

✓ Three types of devices occur in a resistive circuit: (1) resistors, (2) sources, and (3) operational amplifiers.

5.2   Chapter Recap

1. Divide your writing thoughtfully into paragraphs. When a reader sees an indent, he or she will expect it's there for a reason.

2. A declarative sentence needs a subject and a predicate. That is, it must say something about something.

3. Imperative sentences can be used to issue polite commands such as *Let* $x = 2$ or *Assume, for the moment, that* $L = 9$.

4. Extremely long sentences can be hard to read, but a succession of extremely short sentences can sound choppy and childish.

5. A formal, technical document must carry proper punctuation and capitalization throughout.

6. The arrangement of words in a sentence — called syntax — is hugely important to the meaning and effect of the sentence.

7. The active voice can make your writing more vigorous, interesting, direct, and concise. These things are good. (Compare: Your writing can be made more vigorous, interesting, and direct through the use of the active voice.)

8. A familiarity with grammatical terms such as *voice* and *subjunctive mode* can give you access to the contents of grammar and style books such as the classic *The Elements of Style*. This, in turn, can improve your writing dramatically. A quick-and-dirty list of some grammar terms appears in the Quick Reference. Although these terms were once taught in grade school, your own formal education may have omitted them.

9. An internet search box is not an adequate replacement for a good dictionary or thesaurus.

10. Fluff has a way of sneaking into written work. You should revise it out before your reader sees it.

11. Consider parallel construction where appropriate.

12. If you say *bigger*, your reader will ask *bigger than what?* If you say *faster*, your reader will ask *faster than what?* If he or she can't answer such questions, you've used comparatives in the wrong way.

13. If you write the plural pronoun *they* when referring to a singular subject such as *an apple*, then you have a problem with grammatical agreement in number.

14. If you write the pronoun *it* and your reader wonders what *it* refers to, you have an unclear pronoun referent.

15. Try to place modifiers close to the words they modify. The bigger the separation, the bigger the chance for confusion.

16. Don't shift verb tense willy-nilly. You'll be more understandable if you hold to a stable frame of reference.

5.3 Exercises

5.1. Which of the following are declarative sentences? Which are imperatives?

(a) Let I be the electric current flowing through R_1.

(b) This vector can be moved to any other point of the rigid body.

(c) Should we care about accuracy in this case?

(d) Chapter 1: Introduction

(e) Find l.

(f) Design a system to meet all specifications.

(g) The previous two chapters dealt with the analysis and synthesis of linear circuits.

(h) Use the perturbation method to locate a zero of the function f.

(i) Let the starting point be given by $t = 4x/9$.

5.2. Divide each sentence into subject and predicate.

(a) Newton's second law can be stated as $F = ma$.

(b) We define f_0 as the maximum value of $f(x)$ for $x \in [a, b]$.

(c) The first part of this report deals with background theory.

(d) The graph of such a function is obtained as follows.

(e) This challenge we accepted.

5.3. Insert a comma to improve the readability of each sentence.

(a) If each of the outputs is given by (1) it is a simple matter to show that D is given by $D = \sin x$.

(b) With an assumed value of $w = 2$ equations (1) and (2) combine to yield $p = y^2 u$.

(c) As an illustration of the use of (3) first consider the system shown in Figure 2.

(d) When both switches are closed the system operates at a higher speed.

5.4. Insert a pair of commas to mark off the parenthetical phrase.

(a) Resistor R_L the load resistor is chosen next.

(b) The solution for example is not unique.

(c) That particular design regardless of its low cost is unacceptable.

5.5. Fix each run-on sentence.

(a) We treated an analogous system in Chapter 1 the method of solution is the same.

(b) We refer to expressions of this type as waves additional waves are treated later in this section.

(c) To evaluate x, multiply (1) through by y and integrate this gives $x = 2/c$.

(d) Equation (2) is the f-transformation it is closely related to the g-transformation of Chapter 10.

5.6. Criticize the following sentences. Then fix each one if you feel you can.

(a) Digital circuits are used both for communications, computation, and other tasks.

(b) Helical antennas have the advantage of being a robust transmission method.

(c) The addition of additional components does not improve system performance.

(d) Note that the resulting ODEs are homogeneous.

(e) By substituting (2) into equation (3) results in $f = g$.

(f) Now, by introducing $x = a/2$, equation (4) can be written as $f = a^2$.

(g) These are a set of concentric circles.

(h) By factoring out x from the numerator, results in $f = x(y + z)/2$.

(i) The two layers were fastened together with staples.

(j) The process was accelerated by the presence of the new machine.

(k) Our procedure is different than the conventional procedure.

(l) Neither the current nor the voltage are large enough.

(m) There are two general types of friction; static friction and dynamic friction.

(n) Each of the systems were analyzed using the Krauss method.

(o) Due to symmetry, the problem was easily solved.

(p) Where c is the speed of light given by $c = 1/\sqrt{\mu\epsilon}$.

(q) Having simplified expression (4), the conclusion is that $x = 2y/3$.

(r) Results obtained using the FEM method is also provided in the table.

(s) The machine consists, of a given electric motor, with a propeller attached to its shaft.

(t) It should be borne in mind that $F = ma$.

(u) As a new engineer, the safest thing to do is ask a more experienced colleague.

(v) Next, we integrate both sides of equation (1.2) to obtain: $x = at^2/2$.

(w) Therefore, on the surface of the box, S, see Figure 4.2, $h = l^2/9$, where l is the longest edge dimension of the box.

(x) To start with, we check the efficiency at full load.

(y) The top view of the apparatus then looks like shown in Figure 1.

(z) So, finally, we find for the distance $x = at^2/2$.

5.7. Repeat:

(a) The denominator of equation (1.45) may be written as, using equations (1.4) and (1.44), $x^2 - a^2$.

(b) In Figure 2 we show $f(x)$ and $g(x)$ as a function of x.

(c) An expert informed us last week that electric charges always produced electric fields.

(d) We would have preferred to have solved the problem more quickly.

(e) A better optimization method will pave the way for progress.

(f) Then, in Chapter 2, we propose a method of approximation of the radiation fields of the antenna.

(g) The two solutions of the equation $x^2 = 1$ are, $x = 1$ and $x = -1$.

(h) A passive circuit may contain: resistors, capacitors, inductors, and sources.

(i) This is not the exact solution but, it is a good approximation.

5.8. Distinguish between the words in each of the following pairs:

(a) *can* and *may*.

(b) *farther* and *further*.

(c) *imply* and *infer*.

(d) *fewer* and *less*.

(e) *partially* and *partly*.

(f) *in* and *into*.

(g) *proved* and *proven*.

(h) *beside* and *besides*.

(i) *provided* and *providing*.

(j) *generally* and *usually*.

(k) *lay* and *lie*.

(l) *appear* and *seem*.

(m) *compose* and *comprise.*

5.9. Composition rule #14 in Strunk's *Elements of Style* is

Avoid a succession of loose sentences.

What is a *loose sentence* in English? Which of the following sentences is loose:

(a) The machine finally failed today, after four months of overheating.
(b) After four months of overheating, the machine finally failed today.

Give a suggestion for how loose sentences can be avoided.

5.10. Which makes more sense?

We wish it was possible.
We wish it were possible.

Repeat for

The system acted as though it was going to fail.
The system acted as though it were going to fail.

6

Keep Your Reader in Mind

Grammar and style are important; we offered our short list of rules and suggestions in Chapter 5. Here we consider the content of your writing, and urge you to always keep the reader in mind. Make it a goal to ease the burden on the audience. Care invested in a document will be greatly appreciated.

6.1 More Rules and Suggestions

Proofread every sentence. Revise, revise, revise.

Put yourself in the position of the target reader, and read.

Read each piece as you complete it. After each edit, return to the start of the paragraph and read the change in context. If you think you're done writing, read the *entire* manuscript from start to finish. Read for comprehension, assuming you know only what the target reader should be expected to know.

Does each sentence make sense individually? Does every paragraph worth of sentences add up to what is intended? Does every section worth of paragraphs add up to what is intended? Does every chapter worth of sections add up to what is intended? Does the entire document add up to what is intended? If not, revise and repeat.

Then, if possible, have a colleague read and comment.

Make definite assertions. Communicate thoughts.

(a) Rely mostly on simple, declarative sentences.

A typical sentence should say something that could be judged as true or false.

> **Example.** Although noun phrases such as
>
> huge force
>
> optimal system configuration
>
> the Hertz potential expression

represent mental objects, they do not represent thoughts. Even if we adjoin verb forms to these noun phrases, we do not get thoughts:

huge force acting

to seek the optimal system configuration

simplifying the Hertz potential expression

This is why sentence fragments are unacceptable: they do not express thoughts.

A thought involves some relation between mental objects.

Example. Here are some thoughts:

✓ There was a huge force acting on bolt C.

✓ We decided to seek the optimal system configuration.

✓ Simplifying the expression, we obtain a key result.

Each is a thought because something is asserted about something.

(b) Avoid vagueness and ambiguity.

A passage is vague if it is unclear or indefinite.

Example. The following passage is vague.

Precise measurements are taken at short intervals.

Available information was simply left out. Better:

✓ Measurements accurate to 1 nW are taken at 1 μs intervals.

A passage is ambiguous if it has more than one possible meaning.

Example. The following passage is ambiguous.

After that, we fixed the load resistor.

The word *fixed* has more than one possible meaning. Was the load resistor broken and then repaired? Or was it assigned a specific and unchanging value after being permitted to vary?

Monitor your choice of terms.

(a) Use the right technical term.

Word choice is crucial in technical writing. Don't settle for a word whose meaning is merely close. Insist on the right word.

> **Example.** Don't say *resistance* if you mean *resistor*:
>
> The resistance was soldered into place.

A *resistor* is a *device* and a *resistance* is a *numerical value along with a unit*. We cannot solder a numerical value onto a breadboard. Could a reader figure out what is meant? Probably — but remember, that's not his or her primary job in this author/reader interaction. You should *say* what you mean. Use the right technical term.

It is equally important to use nontechnical words correctly.

> **Example.** Don't be the careless writer who chooses the wrong word from a sound-alike pair such as *affect/effect*.
>
> We decided to determine the *affect* of this high voltage value . . .
>
> One must apply the basic *principals* of circuit theory.
>
> All design possibilities were rejected *accept* one.
>
> The failure of this machine was long *overdo*.
>
> Post-consumer *waist* was a concern for management.

> **Example.** Don't use the words *myself*, *yourself*, and *ourselves* in place of *I*, *you*, and *we* as subjects of sentences. This is incorrect:
>
> Bob and myself completed the report last night.

A writer should take the time to care about words. An electronic spell-checker won't save you from having to consult a dictionary from time to time. It won't save you from writing *in the presents of* instead of *in the presence of*. Some groups of words that people often confuse are given on p. 164. But there are *many* of these in English and we display these only to convince you to have a dictionary at hand.

The people around you will often use words imprecisely. As an educated professional writing highly technical material for wide distribution and possible archival value, you cannot afford to do so. Obtain a grammar book and peruse it once in a while. We'll recommend some books later on, but practically any book on English usage will provide many useful warnings about common misuse.

Example. The words *between* and *among* are not interchangeable in English. *Between* is used for two things, while *among* is used for three or more things. The passage

The total charge Q was equally distributed between the three spherical electrodes ...

could definitely be improved.

Appropriate word choice can help you avoid pompous language.

Example.

The water leak had a deleterious effect on the floor.

✓ The water leak stained the floor.

We could go on forever about words, but let's move on.

(b) Define all crucial terms.

Every term employed must be either defined or already understood by the audience. This is especially true for terms on which the exposition hinges. Formal definitions are described in Chapter 7. For informal definitions, many variations in wording are permissible.

Example.

✓ We define the *output section* of our circuit as that section consisting of components D and E together with the output port W.

✓ Therefore $f(x) = \sin x + e^{x^2} - 4\sqrt{ab}$. Writing $g(x) = e^{x^2} - 4\sqrt{ab}$, we have $f(x) = \sin x + g(x)$ and hence ...

✓ The notation $\Re[z]$ stands for the real part of z.

✓ The voltage across R_1 is given by $V_1 = R_1 I_1$, where I_1 is the current through R_1.

We do not adequately define a term by providing synonyms for the term, giving examples of the term, using the term in its own definition, or merely stating things relevant to the term.

Example. The following are not good definitions.

A *force* is a push or a pull.

Metals are things like copper and aluminum.

Bridge design is the practice of designing bridges.

Electric current causes voltage drops and Joule heating in resistive materials.

(c) Use signaling words and helpful phrases.

Assist the reader by putting him or her on notice about your intentions.

Example. When generalizing or particularizing, say so. Look at these passages:

✓ Therefore the equivalent resistance of the series combination is $R_e = R_1 + R_2$. More generally, the formula for N resistors is $R_e = R_1 + R_2 + \cdots + R_N$.

✓ The force acting on the nth structural member is given by $F_n = 2/n$. In particular, the force on the third and crucial member is $F_3 = 2/3$.

Notice how the phrases *more generally* and *in particular* signal you about what to expect from the author, thereby making reading easier.

Other helpful introductory phrases are listed on p. 161. But please employ them sensibly.

Example. Don't write

Conclusion. In conclusion, . . .

This would be quite redundant.

Example. If you write

On one hand, . . .

then later you should write

On the other hand, . . .

In other words, be sure to follow through.

In Chapter 2 we mentioned *signposts*: words that serve as headings, subheadings, or other types of labels to guide the reader. Mathematicians seem to rely mostly on the following:

Claim	Proof	Comment	Problem
Corollary	Proposition	Definition	Question
Example	Remark	Lemma	Solution
Notation	Summary	Note	Theorem

These are great for engineering writing as well, but engineers deal with so many issues that additional signposts suggest themselves. How about *Background*, *Goal*, *Interpretation*, or *Rationale*? Imagine opening a technical book and seeing it peppered with headings of this type. A sense of helpful organization would certainly emerge. A list of some possibilities appears on p. 163.

We also talked about *premise indicators* and *conclusion indicators* before. Let's list some of these for reference.

premise	conclusion
since, because	therefore, hence, thus
in view of the fact that	we conclude that
by virtue of	it follows that
for	consequently
inasmuch as	we may infer that
as indicated by	which implies that

Again, these should only appear in argumentation.

Example. We often see students misuse conclusion indicators. They seem especially prone to use *therefore* when no logical conclusion has been made.

I first connected the wire from A to B. Therefore, I connected another wire to B.

This writer may be confusing *therefore* with *thereafter*.

(d) Search for a good verb.

The English language is rich with verbs. Try to use vivid ones.

Example. Write *simplify* rather than *reduce in complexity*.

Write well-conceived descriptions.

When writing a description, keep the following in mind.

(a) Keep your standpoint clear.

Suppose you are describing the appearance of a machine. You want the reader to see it as you see it. Where are you located as an observer? What you see while *outside* the machine will differ from what you see while *inside*, and what you see while *above* the machine will differ from what you see while *below* or *to the side of* it. Make your descriptive standpoint clear. If you must shift it during the description, be explicit about that.

> **Example A.**
>
> We may investigate a simple magnetic field through the use of iron filings. Suppose we run a long straight wire vertically through a horizontal piece of white cardboard. A battery is connected across the wire in such a way that the resulting electric current flows upward through the wire. By sprinkling some iron filings around the wire and gently tapping the cardboard while observing from above, we see a pattern of circles emerge. The iron filings are then arranged along the lines of magnetic field.

(b) Put great thought into selection of details.

Include what is necessary and no more.

> **Example.** In the description of Example A above, it was not necessary to specify the exact source of the white cardboard: any white cardboard would do. On the other hand, perhaps the battery voltage should have been specified for the success of the demonstration and for the safety of the participants.

(c) Arrange the chosen details well.

Provide the reader with a coherent picture. If certain details are grouped naturally in your subject, they should be grouped intentionally in your writing. Don't jump around haphazardly.

> **Example B.** Contrast the description of Example A with the following.
>
> Iron filings can show us magnetic field lines. A battery will be

needed to produce current. The filings will arrange themselves in circles around a wire. Put them on cardboard and make sure the wire runs vertically. Tap gently while holding the cardboard horizontally. The phenomenon is best observed from a top view. White cardboard should be used.

We hope you'll agree that this piece of writing leaves much to be desired.

Crucial details can be emphasized by placing them first or last in the description.

Example. Let us compare the descriptions of Examples A and B with regard to the order in which things are presented. The description of Example A starts with the central idea: we can *investigate a simple magnetic field*. Iron filings are used to accomplish this, but they are not the whole point of the description. The description of Example B, on the other hand, commences with *iron filings*. Now consider how the two descriptions end. The description of Example A ends with identification of the magnetic field lines: again, the main goal of the demonstration. The description of Example B, on the other hand, ends with the color of the cardboard.

These considerations show how important it can be to *generate a plan* for writing a description. Good descriptions don't just happen — they are deliberately and skillfully *made*.

Decompose a long development into labeled steps.

If an argument spans many pages, perhaps it should be broken into reasonable chunks.

Example.

For convenience, we present the argument as a sequence of steps.

Step 1. Suppose f is a ...

Step 2. Now let ...

Step 3. Finally, ...

This concludes the argument. Note that ...

When writing instructions, treat the steps in the desired order of execution.

Instructions should be written as imperative sentences.

Example.

✓ Carefully but firmly insert the wire into line B of the breadboard.

The wire should be inserted carefully but firmly into line B of the breadboard.

Don't ask the reader to contort mentally while trying to follow instructions.

Example.

Holding the nail with your left hand, carefully pound it into the wood with the hammer after picking up the latter with your right hand.

✓ Holding the nail with your left hand, pick up the hammer with your right hand and carefully pound the nail into the wood.

You may wish to precede instructions with a list of materials that will be needed to execute them.

Form well-conceived classifications.

Engineers are often called upon to classify things (objects, ideas, methods, and so on). A well executed, systematic classification scheme can make sense out of bewildering complexity. A poorly executed scheme, on the other hand, can confuse and mislead those who attempt to rely on it. Follow these pointers when putting forth a classification in a formal document.

(a) Be clear about the universe of discourse.

By *universe of discourse* we mean the set of all things that will be covered by your classification scheme. Should you really try to classify all forms of watercraft (ships, commercial boats, pleasure boats, rafts, canoes, kayaks, surfboards, inner tubes, etc.) or will ships be enough? Should you really try to classify all ships, or will a certain subset (such as those currently in use by a certain country during a certain time period) be enough? The more you limit this initial set of things, the easier the classification task will become. On the other hand, you can't afford to omit things essential to a discussion.

(b) Include everything in the universe of discourse.

Suppose you have decided to classify modern-day Norwegian ships. Then your classification scheme must include *all* modern-day Norwegian ships.

(c) Carefully choose and maintain a principle of classification.

On what basis will you subdivide the set of all modern-day Norwegian ships? Shape? Size? Weight? Function? Color? This question matters very much and could determine the value of the classification scheme to others. In any case, your choice must be carried out consistently. If there are 5000 ships to classify, you cannot put 2000 of these into 20 subclasses by weight and the remaining 3000 into 10 subclasses by function.

(d) Make sure the subclasses do not overlap.

Suppose you decide to classify 5000 ships into 20 subclasses by weight. Then no single ship should fall into more than one subclass. In other words, a classification scheme must be *mutually exclusive* as well as *collectively exhaustive*.

Example. Suppose we need a simple classification of transistors. We could class them structurally as bipolar junction transistors (BJTs) and field effect transistors (FETs). This classification principle is valid because a given transistor must fall into exactly one of these two categories. If we wanted to take the classification further, we could subdivide each category by application as power transistors, switching transistors, and amplifying transistors. The result is the two-level classification scheme shown in Figure 6.1. An engineer who failed to heed the main principles of classification might class all transistors as either BJTs or switching transistors. The problem is that some BJTs are also switching transistors, and some transistors are neither BJTs nor switching transistors. Confusion is sure to result.

Use the standard unit prefixes when appropriate.

The list of unit prefixes (such as *pico, nano, micro, milli, kilo, mega,* and *giga*) appears on p. 161.

Example. Write $V_0 = 765$ nV rather than $V_0 = 0.000000765$ V.

Don't plural-ize unit symbols by adding *s*.

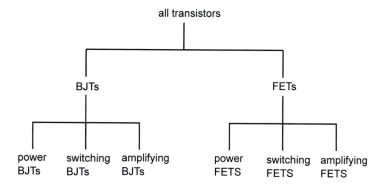

FIGURE 6.1
A simple classification of transistors.

Example. Write 5 m (not 5 ms) for 5 meters.

Older books may use double prefixes. These are no longer acceptable.

Example.

The capacitance is 10 $\mu\mu$F.

✓ The capacitance is 10 pF.

Be careful with computer units.

Example. A *kilobyte* (kB) is generally understood to mean 1000 bytes, whereas a *Kilobyte* (KB) means 1024 bytes.

Some specific combinations of units and prefixes may have dedicated terms. Although these may not be preferable, they are in widespread use. Examples are *micron* for μm, and *mil* for 0.001 inch.

Use analogies but use them with honesty.

Analogies can be helpful; we mentioned this in Chapter 2. We also cautioned that there is a big difference between saying that *an atom is like a tiny solar system* and *an atom is a tiny solar system*. A solar system has a star at its center. An atom (in a very simplistic picture) has a nucleus at its center. They are radically different physical objects, but a mental picture of one can assist — to an extent — with a mental picture of the other one.

Be consistent with aspects of visual format.

(a) Display subdivision heads consistently.

Formal documents are often divided into chapters, sections, and subsections. The corresponding titles should be displayed consistently throughout a document. Don't double-space certain sections and single-space the rest, show some equation numbers at the right margin and others at the left margin, or leave other glaring inconsistencies in a final document.

(b) Use symbolic notation consistently.

The reader will not appreciate an arbitrary re-use of variables. Don't use the Greek letter λ for three different things in the same document without an adequate explanation. Don't expect the reader to figure out that N is the same as N, or that n is the same as N or $\mathbf{N}$.

Include a notation guide for long works.

A list of symbols can help the reader navigate highly mathematical books, dissertations, and reports. You could construct one in various ways, showing symbols, their meanings, their first occurrences in the document, etc.

Example.

Key to Notation

Symbol	Meaning	Page
V	linear space	4
B	Banach space	8
H	Hilbert space	9
W	Sobolev space	15

Consider whether an index is appropriate for a document.

A book certainly needs an index; other long documents may benefit from one as well. Here are some pointers on making one.

(a) Alphabetize the index.

An index in which references are arranged in order of appearance in the book is really just a fancy table of contents.

(b) Distribute references under sensible head words.

If your book has a section called *The Transistor*, then the corresponding index reference should be *transistor* or *transistor(s)*. Nobody wants to use an index where one must search for words under the English articles *the*, *a*, or *an*. If your book has a section called *On Power Systems*, the corresponding index entry should be *power systems*.

(c) Include all subjects in the index.

If your book mentions antennas and specifically dipole antennas, create index entries such as

> antenna(s), 53
>   dipole, 54
>   ⋮
> dipole antenna, 54

If it's worth including in a book, it's worth including in the index.

(d) Include all occurrences of a given subject in the index.

If the dipole antenna is also mentioned on pages 78, 89, and 95 of your book, then provide this information:

> antenna(s), 53
>   dipole, 54, 78, 89, 95
>   ⋮
> dipole antenna, 54, 78, 89, 95

If it's worth including in a book four times, it's worth including in the index four times.

Index creation is greatly facilitated by computer software and its ability to implement automatic, dynamic cross-referencing. Creation of an index entry in a LaTeX document is addressed in Exercise 7.5.

Don't annoy the audience on a personal level.

(a) Avoid the first person singular in formal, technical writing.

Unless you're really famous in your field and writing your memoir, you'll probably want to avoid *I* and *me* most of the time.

Example. This is first person:

> I remember the day I made the big discovery. All at once it dawned
> on me that if I substituted ...

Mathematical writing, already heavily technical, often relies on the second
person singular (*we* and *us*) to keep the reader personally involved.

Example. Mathematicians, along with many physicists, write like this:

> Let us define a quadratic function $q(x) = ax^2 + bx + c$, where a, b, c
> are constants. Taking the derivative, we obtain ...

There is normally no reason why an engineer should be prohibited from
doing so, when writing down mathematical arguments.

However, some technical journals (such as the *IEEE Transactions on Education*) insist on the third person.

Example. This is third person:

> Hence the existence of at least one solution is assured. This solution
> is sufficient for practical purposes if it is well-behaved (i.e., if it is
> continuous with a continuous first derivative).

There is no mention of *I*, *me*, *we*, or *us*.

(b) Know the audience's stance on gender-neutral language.

Gender-neutral language entails the use of constructions such as *he/she*, *s/he*,
and *him/her* to avoid *he*, *she*, *him*, and *her*.

> The engineer is wise to supplement his or her knowledge through
> constant study of technical materials.

Pluralization is sometimes recommended as a way to avoid gender oriented
language.

> Engineers are wise to supplement their knowledge through con-
> stant study of technical materials.

Unfortunately, these approaches can lead to constructions that are stylistically awkward or even grammatically incorrect. Our preference is for a conscious alternation between the male and female, especially for use with examples in instructional materials such as textbooks.

Example.

> *Example 1.* For her final circuit design, Sally decided ...
>
> ⋮
>
> *Example 2.* Robert, a civil engineer, approached his project ...
>
> ⋮
>
> *Example 3.* Mary implemented her subsystem using ...
>
> ⋮

Your preference may differ. Moreover, you may find yourself preparing a document for a group, institution, or agency that requires gender-neutral language for all submissions. See, for example, Exercise 6.10.

(c) Avoid colloquialisms, idioms, and inappropriate informality.

These things can be bad for international communications and for communication across generations.

Example. You might say

> ... and therefore the second term on the right-hand side goes away

in a personal meeting with colleagues, but don't write it. You might refer to mathematical substitution as *plugging in*, but don't write it.

A short table of suggested replacements appears on p. 165.

Don't get fancy.

(a) Don't coin new words when existing words will do.

Example.

✓ Circuit number 4 radiates strongly in the z-direction.

Circuit number 4 is strongly radiational in the z-direction.

(b) Be careful with acronyms.

Unless an acronym should be familiar to the audience, expand it on first use.

Example.

✓ We will use the Fast Fourier Transform (FFT) to accelerate the computation.

We will use the FFT (Fast Fourier Transform) to accelerate the computation.

The use of acronyms cannot always be recommended. Acronyms are appropriate when widely used or even expected by certain audiences. Communication engineers, for instance, expect the use of AM for *amplitude modulation*. Acronyms are also appropriate for long, important terms that occur frequently in a document. Don't coin an acronym for a relatively unimportant term or one that seldom appears. Acronyms should be used as conveniences for the reader. They shouldn't be invented because the writer thinks they "sound cool." See also Exercises 6.13 and 6.14.

(c) Use only standard abbreviations.

Some standard abbreviations used in English writing were given on p. 70. In formal writing, do not use nonstandard abbreviations such as

w.l.o.g. — without loss of generality
s.t. — such that
w.r.t. — with respect to

The abbreviation *iff* (*if and only if*) is a borderline case. It is common in mathematical writing but should be defined in a document meant for a broad engineering audience.

(d) Never use internet slang, phrases, or abbreviations.

Cute terms like *eddress*, *grats* and *IMHO* don't belong in technical documents.

(e) Don't use simplified spelling.

Avoid *thru*, *tho*, *nite*, and *til*. Certain illiteracies are becoming more common within technical jargon, even in material published by equipment manufacturers. Spell words conventionally in formal documents.

Example. There is a calibration technique known as the *thru-reflect-line method*. This does not give us license to write *thru* instead of *through* for other purposes.

(f) Don't mix British and American spelling patterns.

The British spelling is *colour* but the American spelling is *color*. Further examples are given on p. 163. Consider the target audience, and be consistent.

(g) Understand non-English expressions if compelled to use them.

Don't guess at the meanings of terms such *a fortiori* and *ad hoc*. Our advice is to avoid them, but those who insist on using them should be sure of their meanings. A brief reference list appears on p. 166.

> **Example.** The Latin *quod erat demonstrandum*, abbreviated Q.E.D., means *which was to be demonstrated*. It would be redundant to write
>
> ... and hence $v = at$, which was to be demonstrated.   Q.E.D.

(h) Refrain from rhetorical questions.

These are things like

> Isn't this a strange phenomenon?
>
> Why is that?
>
> What can be done in such a case?
>
> How did that ever happen? Are they joking? Are you sure?

to which answers are neither needed nor expected.

Avoid things that are too long.

(a) Avoid long noun phrases.

> **Example.**
>
> The large black silicon parallel safety overload diode $D17$ is optional.
>
> annoyingly long noun phrase

(b) Avoid lengthy parentheticals.

A long string of words in parentheses can cause the reader to lose his or her place in a sentence.

Example.

> Assuming that $g(x)$ is continuous (we should recall that, by def-
> inition, a real-valued function f of a real variable x is continu-
> ous at $x = x_0$ if for every $\varepsilon > 0$ there exists $\delta > 0$ such that
> $|f(x) - f(x_0)| < \varepsilon$ whenever $|x - x_0| < \delta$), we can apply Theo-
> rem 1.2.

Better:

> Recall that, by definition, a real-valued function f of a real variable
> x is continuous at $x = x_0$ if for every $\varepsilon > 0$ there exists $\delta > 0$ such
> that $|f(x) - f(x_0)| < \varepsilon$ whenever $|x - x_0| < \delta$. Assuming $g(x)$ is
> continuous, we can apply Theorem 1.2.

(c) Avoid unnecessary repetition of words.

Repetition of the same word or phrase over a short span can be especially
irritating to readers.

Example.

> Transistors are characterized by a number of parameters and can
> be classified into a number of different categories. In the selection
> of transistors, a number of technical parameters must be specified.
> These essential parameters are reviewed in Chapter 2.

In this case, *parameters* and *number* each appears three times within a
span of three sentences.

(d) Remove the scaffolding when the work is done.

During the writing process, we often have thoughts like

> Having finished that topic, the next thing I need to discuss is . . .

These can be regarded as forms of scaffolding necessary for us to build our
written piece. Sometimes we even write them down. However, the reader may
not benefit from scaffolding left behind. A similar idea applies to repeated
equations. Suppose that on page 465 of a manuscript you must refer back
to equation (2.103). If that equation occurred on page 32 of the manuscript,
perhaps it should be repeated for the reader's convenience (space permitting,
of course). However, if it occurred on page 463, perhaps it should not be
repeated in the final document — even if you must temporarily repeat it for
your own convenience during the writing process.

Pay attention to what you're saying.

(a) Use sensible variable names.

Consider which letter is best for a given purpose. The Greek alphabet is summarized on p. 159.

Example. If you have a choice, a time-related quantity might be better symbolized as t, T, or τ than as p, L, or ψ.

(b) Explain all diagrams.

A common student error is to include diagrams or figures without context or meaning. Every figure included must be discussed, or at least mentioned, in the text.

Explain everything that appears in each diagram. Don't leave the reader scratching his or her head about what that black triangle represents or what that little arrow means. Examine the diagram, pretend you're encountering it for the first time, and then systematically explain it. But insert the explanation in the body of the text, not in the caption. Excessive detail makes captions annoyingly long, especially if repeated across a sequence of figures.

By the way, the terms *plot* and *graph* are not always permissible substitutes for the term *figure*. A diagram of a machine should not be called a *plot* or a *graph*. Plots and graphs depict mathematical data, relations, and results.

(c) Don't write nonsense.

Nonsense should never appear in engineering writing. There are potentially so many types of nonsense that we could not possibly make a reasonable catalog here. A few examples will have to suffice.

Example. A real number, by itself and without further context, cannot be called *large* or *small*. Is 1×10^{-9} a small number? No, not when compared to 1×10^{-37}. Is 10^{428} a large number? Not when compared to 1000^{428}. The real numbers go on forever in both directions, positive and negative. Not even zero can be called *small*, as -1000^{45} is quite a bit smaller. (It's ok, however, to say that zero is *small in absolute value*.)

Example. Consider the English abomination *and/or*.

We decided to include a safety valve and/or an emergency stop button.

What precisely does this mean? What did they decide? The word *or* would work much better here, as in its normal "inclusive" meaning it also covers the meaning of *and*. Another abomination, at least in technical writing, is *more or less*.

The mechanism was more or less destroyed.

What does this mean? Another one is *or so*.

The output power was 10 Watts or so.

What does this mean? If P is the output power in Watts and $9 \leq P \leq 11$, can we say that P is 10 *Watts or so*? How about $2 \leq P \leq 18$?

(d) Refrain from lazy, indefensible assertions.

An engineering document is no place for bluffing.

Example.

It should be noted that no author has explored ...

No author has explored that? Where is the evidence?

... all possible modes were considered ...

All possible modes? Can this author prove that only seven modes exist and that he or she considered them all?

In recent years there has been increasing interest in ...

There has? Who says?

Recently, there has been great interest in ...

Interest by whom? The author? Some identifiable, quantifiable community of investigators? Has the author done a survey?

It is believed that ...

It is well known that ...

By whom? The author? The general community of people who know about such things? Everyone on the planet? Someone somewhere?

There is reason to believe that ...

Maybe the reason should be indicated.

We conjecture that ...

Is the author just guessing?

Similarly, be careful with the terms *minimal* or *maximal* unless you can back them up (e.g., unless you can prove you've found a min or a max by equating a first derivative to zero and then following up with the second derivative test).

Example.

We believe we have found the ideal solution to this problem.

Under what circumstances should one label something as *ideal* in technical writing? How would you back up an assertion that a solution is ideal?

What exactly is meant (technically speaking) by terms such as *very large, fairly large, rather accurate,* and *somewhat inaccurate?* What is meant by *substantial, considerable,* or *significant?* What is meant by *good agreement?*

Be careful with the phrases *it is seen that* and *it is clear that.* Is it reasonably certain that the reader will see what you see? Consider explaining the topic thoroughly rather than relying on these presumptive phrases.

Example. Also be careful with saying that things are *obvious.*

It is obvious that no other solution exists.

It is? Has uniqueness been established? (This issue will be discussed on p. 136.) In other words, can the assertion be defended?

Keep the visual aspects simple.

(a) Know that typography is tricky.

Things like typeface selection and page layout should often be left to the experts. When in doubt, try to keep it simple. The visual design of a document should not call attention to itself. Its sole purpose is to support clear communication.

Example. Don't go crazy with typefaces or fonts.

The output voltage got *larger* and **larger** and LARGER.

Will this help the reader?

The availability of many visual options does not imply that they should all be used.

Example. Use a minimum 6-point font size for figure callouts. A callout is text used to label something in a figure. Your computer drawing program may supply 4-point fonts, but don't use them (even for subscripts on other symbols).

One way to handle the general layout issue is to use LaTeX, a document preparation system we will discuss on p. 121. LaTeX comes equipped with some standard "document classes" called *article*, *report*, and *book*; these impose reasonable page layouts and other typographical structures so that you can concentrate on the technical content of a writing project. Many book and journal publishers supply authors with special LaTeX *style files* or *class files* to implement their preferred layout characteristics. This is one way of getting professional layout assistance for free.

(b) Avoid superfluous clutter.

Again, keep things simple whenever possible. Your subject is probably complicated enough as it is.

Example. Consider:

$$x = \left(\frac{b}{a}\right) \qquad \text{and} \qquad x = 2 \cdot \pi \cdot r$$

In the first equation, why do we need parentheses on the right-hand side? In the second equation, why do we need the centered dots? Is our reader so unprepared mathematically that we cannot write $x = 2\pi r$?

Inexperienced writers clutter their documents in many ways.

Example. Don't use underlining (an approach left over from the age of the typewriter) to emphasize text.

It is essential to check every connection.

Use *italic* or **bold** for emphasis instead.

6.2   Chapter Recap

1. Proofread, revise, and repeat. If possible, get feedback from a trusted associate.

2. Write to inform.

3. Words mean things in engineering. Know what they mean. *Equation* does not mean *expression*. Choose words with care.

4. Define all crucial terms, especially those that may be unfamiliar to the reader.

5. Do things that will ease the reader's job. Many such things are available, but central considerations are formality, convention, and consistency.

6. Don't write anything you can't support with evidence.

7. Try for a clean, sensible page layout. When in doubt about typographical issues, rely on experts.

6.3 Exercises

6.1. Define the terms *metaphor* and *simile*.

6.2. Distinguish between the technical terms *precision* and *accuracy*. Give examples of correct usage for each.

6.3. List some of the general bases you could cover in describing an engineering system.

6.4. Fix each sentence. Some of the errors present may have been discussed in Chapter 5.

(a) The machine has a stationery element and a moving element.

(b) The principal of operation is simple.

(c) Having found the solution, let us precede to the next step.

(d) Before we can prove the theorem, we must first establish an auxiliary proposition.

(e) The sampling apparatus was lowered down into the tank.

(f) Outside the nozzle, the two flows combine together.

(g) First, the powder must be dissolved in boiling hot water.

(h) We cannot agree with those kind of statements.

(i) Two negative signs cancel one another.

(j) Resistors are more useful than all the circuit elements.

(k) Every engineer should educate themselves in higher mathematics.

(l) The analysis has been finished this semester.

(m) By observing these rules, errors may be avoided.

(n) Mathematics are an interesting and useful subject.

(o) A copy of his course syllabi was requested.

(p) It seems that everybody wanted to express their criticism of the design.

(q) The output voltage was plenty large enough.
(r) A large and a small force was applied.
(s) A large and small force was applied.

6.5. Repeat:

(a) The force was a necessary and a sufficient one for the purpose.
(b) We would fix the machine, if it was possible.
(c) What use is the study of calculus?
(d) The geartrain appears to have reached the end of their service life.
(e) We could not account for all the phenomena that was observed.
(f) The external force was .56 N.
(g) The machine was working at the time.
(h) Each and every real value of x satisfies $x^2 \geq 0$.
(i) Circuit analysis deals with devices such as resistors, transistors, capacitors, etc.
(j) Civil engineers have made great progress in the last two hundred years.

6.6. Consult a dictionary to check your understanding of the following mathematical signposts: (a) lemma, (b) theorem, (c) corollary.

6.7. Use the standard unit prefixes to write each of the following numerical answers in a simpler form:

(a) $R = 100000 \ \Omega$
(b) $I = 0.000000000005$ A

6.8. Criticize the following.

(a) By substituting $x = 5$ into $y = x^2$, I get $y = 25$.
(b) As virtually everyone knows, the UYELYG method does not have the same limitations.
(c) The power company's poured concrete utility room floor drain was plugged.
(d) We have $x = 4$ and/or $x = -4$.
(e) The maximum value was found by a trial-and-error approach.
(f) Similarly x is found via the following relation $x = a^2$.
(g) For a detailed account of some practical issues in circuit design, the reader may look up Smith [2].
(h) In this particular case, the second term drops out.
(i) We made up a new method of solution.
(j) The solution is easier to find when $n =$ odd integer.
(k) The larger this number is, the better off you will be.
(l) We see that the polarization of the electric field is the same.
(m) The cost of such systems is, of course, quite high but improvements, with great success, have been made with respect to frequency performance.
(n) This method permits us to solve problems that could not be solved in any other way.
(o) It is seen that the resulting error is significant.
(p) By substituting $a = b = 1$ gives $f(a, b) = a^2(b + 1) = 2$.

(q) Given a receiver with a bandwidth of 100 kHz and an ordinary monopole antenna.

(r) To correct the problem, we installed the transistor and resistor shown in Figure 2.

(s) Upon overheating, the part took on a green color.

(t) The failure process took place inside of 20 μs.

(u) Old-fashioned analysis is not without its uses.

(v) The system output power was in the neighborhood of 10 Watts.

(w) The mechanism was pretty much outdated.

(x) The comparison that the writer made of those two topics were helpful for understanding.

(y) The process ran its course in 10 mS.

6.9. What is *irony*? Comment on the use of irony in formal technical writing.

6.10. Download the *IEEE Standards Style Manual* and read the section on gender-neutral language.

6.11. Lazy writers sometimes coin new words by *verb-alizing* nouns. Suppose a writer states that he *foiled* an algebraic expression, meaning that he applied the *foil* (*first-outside-inside-last*) method to expand $(a + b)(c + d)$ as $ab + ad + bc + bd$. Comment on this use of the word *foiled*.

6.12. Comment on the use of the term *wattage* for *power*, and *amperage* for *current*.

6.13. A technical journal may have its own policies regarding acronyms. As an example, find the policy implemented by the editorial board of the *IEEE Transactions on Microwave Theory and Techniques*.

6.14. Is it proper to define an acronym in the abstract of a paper and then use it in the body of the paper without defining it again? Why or why not? Is it proper to define an acronym in the main body of a paper but use it in the abstract without defining it there? Why or why not?

6.15. Comment on the issue of possession by inanimate objects. Consider, for example,

... demonstrating the circuit's ability to stabilize the system's output.

Would you prefer to read much material written in this style?

7

Write Your Math Well

Properly formulated mathematical arguments are as important in a technical document as correct spelling, good grammar, and well supported claims.

7.1 What's Wrong with My Math?

Engineers take pride in their mathematical abilities. Many of us chose engineering because we were told at an early age that people who are good at math become engineers. But are we really as good as we think?

As engineers we use mathematical tools to develop our ideas and attain our design goals. It shouldn't be necessary to mention the importance of *doing* math correctly — mathematical blunders can lead to dramatic and disastrous results. In 1907 the Quebec railway bridge collapsed because engineers incorrectly calculated the expected load on the bridge. In 1999 the $500 million Mars Climate Orbiter was lost because the computer software on the ground was designed to send instructions to the spacecraft using imperial units, while the software on the spacecraft required instructions in metric units.

So, we tend to think we are good at *doing* math because we get good results. But are we also good at *writing* math? Engineers use math to communicate with other engineers and to archive their thoughts and ideas. When that communication fails, time and money are lost. What often goes unappreciated is that *to do math correctly, we must write math correctly.*

> **Example.** While designing a new electronic device, Connie needs to calculate the power density at various points to determine the thermal load. She is directed to a report written by Tom, a former engineer at her company, and quickly finds an expression that looks promising. The report is based on Tom's graduate thesis, but does not include the mathematical derivation of the expression she is interested in. No matter, Connie programs the formula into MATLAB®:
>
> $$P(x) = 2.95 \cos\left(202\pi\sqrt{(2h)^2 - x^2}\right) \text{ W/m}^3.$$
>
> The report says that h is the height of the device and x is the distance from

the input terminal, both in meters. Since Connie's device has a height of 1.5 mm and she is most interested in the power density at the output, she chooses $h = 1.5 \times 10^{-3}$ m and $x = 10 \times 10^{-3}$ m. This yields an improbable value

$$P = 628 \text{ W/m}^3 .$$

Connie knows this can't possibly be correct. What went wrong?

Tom expressed the power density as a formula for $P(x)$. It is not enough to provide a formula, however; the domain of the function must also be specified. Unfortunately, Tom neglected to state that his formula holds only when the distance from the input terminal is less than or equal to the height of the device. Hence the domain of $P(x)$ is described by

$$0 \leq x \leq h .$$

When Connie took a number well outside this interval, the argument of the square root became negative, the value of the square root function became imaginary, and MATLAB computed the result using the hyperbolic cosine, which grows exponentially with argument and yields a ridiculous result. Connie is perplexed until she finds a copy of Tom's thesis and learns of the restriction on x. Tom failed to communicate his mathematics properly. His failure to include the domain of the function caused Connie much consternation and loss of time.

This was a case of faulty communication caused by a failure to *write* math correctly. Now let's look at an example of a failure to *do* math correctly.

Example. Alexis has designed a system with the input/output relationship

$$y(x) = 2x^2 + 8x ,$$

where x is the input in mV and y is the output in mV. The input is restricted by the system's electronics to the interval

$$-10 \leq x \leq 10 \text{ mV}.$$

The output is fed into a sensitive amplifier that will be destroyed if its input is negative. In order to determine the range of x needed to keep $y(x)$ positive, Alexis reasons as follows.

Starting point:	$2x^2 + 8x > 0.$
Divide through by 2:	$x^2 + 4x > 0.$
Transpose the term $4x$:	$x^2 > -4x.$
Divide through by x:	$x > -4.$

She records these steps in her thesis along with an assurance that the amplifier will be safe provided the input voltage to her system exceeds -4 mV. She takes many measurements with input voltages in the range 1–10 mV and everything works perfectly. Alexis graduates with her master's degree. A year later Phil, a new student, inherits Alexis' experiment and refers to her thesis for instructions on how to use the system. Confident when he sets the input at -1 mV, Phil is horrified when the amplifier fails in a puff of smoke. His advisor is livid and won't accept excuses. What went wrong?

This is a case of faulty mathematical argument rather than faulty communication. Alexis concluded that

$$y(x) > 0 \text{ for all } x > -4 \, .$$

But it is clear that
$$y(-1) = -6 \, ,$$

hence her argument is invalid by counterexample. Where did Alexis go wrong? When she divided by x, she should have remembered that if $x < 0$ then the direction of the inequality must be reversed. A proper conclusion is that
$$y(x) > 0 \text{ for } x < -4 \text{ and } x > 0 \, .$$

Alexis might have caught her error by testing her conclusion with a few substitution instances or by simply plotting her function. In any case, her faulty math caused considerable difficulty within her group.

Now we see that to do math correctly, we must write math correctly. Had Alexis written a proper mathematical argument, she wouldn't have drawn the wrong conclusion about her system's input range. To answer the question we posed earlier,

We are not good at doing math if we cannot write math well.

In this chapter we provide a quick overview on how to write math well, including how to formulate a proper mathematical argument.

7.2 Getting Started

Math is, in part, a written language. It requires not only syntax and vocabulary, but also a means of visual presentation. So we start with a few important things regarding notation and layout that time may have dimmed in your memory.

Be careful with terminology.

Example. *Solve* does not mean *evaluate*.

Hence $x = y - 5$. Solving at $y = 6$, we get $x = 1$.

So $x^2 = 4$. Evaluating, we find that $x = \pm 2$.

Both of these are incorrect (interchange the roles of *solve* and *evaluate* to repair them).

Similarly, keep in mind the difference between the terms *function, expression, equation, identity,* and *graph*. These are not synonymous in mathematics. Let us review.

1. A *function* is a single-valued relation[1] between two sets. A function f from a set X to a set Y is a correspondence between X and Y such that to each $x \in X$ there corresponds a unique $y \in Y$. We write $y = f(x)$. A function consists of a domain, a range, and a mapping rule.

For instance, the equation

$$f(x) = 2x + 1$$

defines f as a linear function of x on the interval $[0, 1]$ (the domain of f). The range of this function f is the interval $[1, 3]$.

2. An *expression* is anything like

$$x^2 , \qquad \frac{1}{y} \sin x , \qquad \sqrt{x^2 + y^2} , \qquad x^{y^z} , \qquad \cdots$$

You could make the expression x^2 into a function by specifying a domain X such as the entire real line. By specifying X as the closed interval $[0, 1]$ instead, you get a *different* function. But not every function can be written down in terms of a nice expression. (You could easily draw something on graph paper that satisfies the vertical line test but defies description by a simple expression.)

3. An *equation* is formed by putting an equals sign between two expressions:

$$\frac{1}{y} \sin x = \sqrt{x^2 + y^2} .$$

---

[1] In mathematics, a *relation* is defined as a set of ordered pairs. This level of technicality is unnecessary for many engineers. However, the engineer who works with functions a lot may benefit from exposure to a book on set theory, especially if he must define functions as part of his work.

Whether this equation has any solutions might be an interesting question. However, by itself, it does not specify a function. We know that the equation

$$x^2 + y^2 = a^2,$$

where x, y are real variables and a is a nonzero real constant, can specify y as a real-valued function of x on the interval $[-a, a]$ only if we choose one of the two branches of the square root function (remember the vertical line test and see Figure 7.1). Even the equation

$$f(x) = x^2$$

does not fully specify a function until a domain set has been specified.[2]

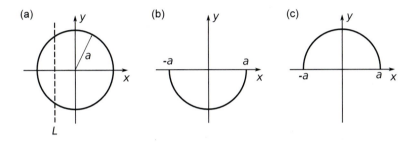

FIGURE 7.1
The terms *equation* and *function* are not synonymous. (a) A circle of radius a in the xy-plane, corresponding to the *equation* $x^2 + y^2 = a^2$. The vertical line L intersects the graph in more than one point. (b) The lower-branch function $y = -(a^2 - x^2)^{1/2}$. (c) The upper-branch function $y = (a^2 - x^2)^{1/2}$.

4. An *identity* in one variable is an equation that holds for all values of the variable. The equations

$$\sin 2x = 2 \sin x \cos x , \qquad x^2 + 2x = x(x + 2)$$

for example, are identities; each holds for *all* real values of x. The equation

$$e^x = 3$$

is not an identity; it holds only for the value $x = \ln 3$.

5. A *graph* is a set of points depicted in two- or three-dimensional space. It

---

[2]However, many mathematicians employ the convention that, unless otherwise stated, f should be assigned the largest domain for which it makes sense. In this case, $f(x) = x^2$ makes sense for any real x. So in a book on real variables, the domain of f would be taken to be the real line. In a book on complex variables, $f(z) = z^2$ would make sense over the whole complex plane. As an engineer applying mathematics to your work, you should keep these sorts of things in mind (and make your own conventions clear in your writing).

may or may not represent a function, and it may or may not be representable in the form of a convenient expression or derivable from a nice equation.

Many other terms are commonly misused.

Example. Another misused word is *arbitrary*. In mathematics, *arbitrary* means *any*. The imperatives

Let ε be an arbitrary positive number.

Let ε be any positive number.

Choose an arbitrary $\varepsilon > 0$.

Choose any $\varepsilon > 0$.

Choose $\varepsilon > 0$.

mean the same thing.

Example. Don't apply the word *arbitrary* to nonuniform samples of a quantity.

✓ The input impedance was measured at chosen frequencies of 2, 3, 5 and 7.7 GHz.

The input impedance was measured at the arbitrarily chosen frequencies 2, 3, 5 and 7.7 GHz.

The second sentence is nonsense. *Arbitrarily* does not mean *randomly* or *impulsively*.

Another misused word is *convergence*. Be careful with asserting that an iterative numerical algorithm *converged* just because you're no longer seeing changes in its output. Many engineers rely on computers nowadays, and they like to assert that algorithms *converge*. But the word *converge* is a reserved word in mathematics. Be careful how you use it.

You examine a sequence of output values from a numerical algorithm and find 1.450, 1.453, 1.454, 1.454. You conclude that the sequence has converged.

Are you sure?

Later you retest the algorithm and follow the sequence a bit farther. You find 1.450, 1.453, 1.454, 1.454, 1.454, 1.455, 1.457, 1.465, 1.489. Now what is your conclusion?

Be sure you can back up a claim of convergence.

Use standard mathematical notation and formatting.

(a) Adhere to standard mathematical notation.

Much of mathematical notation is standardized. It is not advisable to use a symbol such as $*$ or $\bullet$ to denote ordinary multiplication. The same holds for differentiation, integration, and many other operations studied in math courses and applied in engineering courses.

Example.

We have $d = k \star r \star t$ where r is rate, t is time, k is a constant, and $\star$ denotes multiplication.

✓ We have $d = krt$ where r is rate, t is time, and k is a constant.

In cases where more than one notation is acceptable, try to be aware of what is "standard" in your community. For instance, papers in a mechanical engineering journal may regularly use ∂_t for the time derivative, whereas readers of an electrical engineering journal may be expecting to see $\partial/\partial t$. Similarly, in the United States the curl of a vector is written as $\nabla \times \mathbf{A}$ whereas in England it is often written as $\operatorname{curl} \mathbf{A}$.

Your computer system may provide access to symbols that show a superficial resemblance. Choose the right symbol.

Example. Consider the symbols ϵ and $\in$. The first one (ϵ) is a Greek letter, while the second one ($\in$) denotes set membership.

$\alpha = \beta + \in_0$

✓ $\alpha = \beta + \epsilon_0$

$x \epsilon \mathbb{R}$

✓ $x \in \mathbb{R}$

(b) When possible, align your equals signs (vertically or horizontally).

Example.

$$(a + b + c)^2 = (a + b + c)(a + b + c)$$
$$= a^2 + b^2 + c^2 + 2ab + 2bc + 2ca$$
$$= a^2 + b^2 + c^2 + 2(ab + bc + ca)$$

Note the vertical alignment of equals signs in this equation display.

(c) If you must break an equation, do so at a sensible spot.

The best place to break a long equation is immediately before a binary operator such as "+" or "−".

Example. This equation will not fit on one line:

$$\tilde{\mathbf{E}}(\mathbf{r}, \omega) = \int_{V+V_m} \left(-\tilde{\mathbf{J}}_m^i \times \nabla' G + \frac{\tilde{\rho}^i}{\tilde{\epsilon}^c} \nabla' G - j\omega\tilde{\mu}\tilde{\mathbf{J}}^i G \right) dV'$$
$$+ \sum_{n \neq m} \int_{S_n} \left[(\hat{\mathbf{n}}' \times \tilde{\mathbf{E}}) \times \nabla' G + (\hat{\mathbf{n}}' \cdot \tilde{\mathbf{E}})\nabla' G - j\omega\tilde{\mu}(\hat{\mathbf{n}}' \times \tilde{\mathbf{H}})G \right] dS'$$
$$- \sum_{n \neq m} \frac{1}{j\omega\tilde{\epsilon}^c} \oint_{\Gamma_{na}+\Gamma_{nb}} (\mathbf{dl}' \cdot \tilde{\mathbf{H}})\nabla' G, \qquad \mathbf{r} \in V + V_m.$$

However, it is clear that the right-hand side is a sum of three big terms.

(d) Typeset fractions and decimals with care.

When possible, use the solidus (i.e., the slash) for typesetting inline fractions.

Example.

Simplifying, we obtain $x = \frac{b-y}{b+y}$.

✓ Simplifying, we obtain $x = (b - y)/(b + y)$.

Of course, the numerator and denominator of a fraction should be both be centered on the fraction line.

Example.

$$x = \frac{1}{b+y}$$

$$\checkmark \quad x = \frac{1}{b+y}$$

Size a displayed numerical fraction according to the size of the next symbol.

Example.

$$\frac{1}{2}\varepsilon \frac{df}{dx} , \quad \frac{3}{4}\sin\frac{x+y}{x-y} , \quad \frac{1}{2}\int \frac{dx}{x} , \quad \frac{9}{8}\sqrt{\frac{17}{\gamma}} .$$

$$\checkmark \quad \frac{1}{2}\varepsilon\frac{df}{dx} , \quad \frac{3}{4}\sin\frac{x+y}{x-y} , \quad \frac{1}{2}\int \frac{dx}{x} , \quad \frac{9}{8}\sqrt{\frac{17}{\gamma}} .$$

When a decimal number is less than one in absolute value, include a zero before the decimal point.

Example.

The input and output values were .354 and $-.811$, respectively.

$\checkmark$ The input and output values were 0.354 and -0.811, respectively.

(e) Punctuate displayed equations.

In typesetting, an equation such as $F = ma$ is called an *inline equation*, while

$$F = ma$$

is called a *displayed equation*. It's easy to remember to punctuate in-line equations properly, as the following example shows.

Example.

The key is to apply Newton's second law, which we write as $F = ma$. A free-body diagram appears in Figure 2.

Unless told otherwise by an authority figure in charge of your writing project (such as a book editor or a thesis advisor), you should punctuate normally even in the presence of displayed equations.

Example. Consider

> The key is to apply Newton's second law, which we write as
>
> $$F = ma \ .$$
>
> A free-body diagram appears in Figure 2.

Note that the equation display ended with a period.

(f) Use italics consistently (and correctly).

Use italics for variables, not for physical units.

Example.

✓ With $m = 2$ and $a = 4$, we have $f = ma = 8$ N.

  With $m = 2$ and $a = 4$, we have $f = ma = 8\ N$.

Don't use italics for "log-like" functions and operators. These include the trigonometric functions and the symbols for taking maxima, minima, real parts, imaginary parts, limits, and determinants.

Example.

✓ $y = \ln x$, $z = \lim_{n \to \infty} \tan f_n$, $t_0 = \max_k s_k$.

  $y = ln x$, $z = lim_{n \to \infty} tan f_n$, $t_0 = max_k s_k$.

(g) Keep computer variables and computer operation symbols out of mathematical writing.

By *computer variables* we mean things like

```
Distance_Traveled
Acceleration_gravity
Elapsed_Time
```

that might appear in MATLAB or C code. In a formal document, you should define corresponding symbols such as d, g, and t, respectively. Rather than writing

```
Distance_Traveled = 0.5*Acceleration_gravity*Elapsed_Time^2
```

you should write

$$d = \tfrac{1}{2}gt^2 \, .$$

Note that in mathematical writing we do not use the asterisk ("*") to denote multiplication or the caret (as in "x^2") to denote exponentiation.

(h) Don't abuse the mathematical equals sign.

Keep the equals sign out of non-formula text.

Example.

  ... where $d =$ distance between the two points, and ...

✓ ... where d is the distance between the two points, and ...

Your college lecturers took little shortcuts on the chalkboard, but they probably didn't do so in their formal documents.

Never use the equals sign to denote logical implication.

Example. This is really bad technical writing:

  We find that $y = 3$, therefore

$$x = \sqrt{y^2 + 1} \qquad \mathbf{=} \qquad x^2 = y^2 + 1$$
$$\mathbf{=} \qquad x^2 = 3^2 + 1$$
$$\mathbf{=} \qquad z = x^2 = 10$$

We bolded the equals signs that should have been implication arrows.

Similarly, the equals sign cannot be used as shorthand for *the next step in my development is*. That's not what it means!

  Finally, be sure to include equals signs where they are required to make statements.

Example. This is *not* a mathematical argument:

$$(x + 1)^2$$
$$(x + 1)(x + 1)$$
$$x^2 + x + x + 1$$
$$\therefore \ x^2 + 2x + 1$$

In short, the equals sign is important. Use it where it belongs and nowhere else.

(i) Stay with the traditional hierarchy of mathematical grouping symbols.

Use (curved) parentheses first, then (square) brackets if needed, then (curly) braces if needed. So the traditional order looks like

$$\left\{ \cdots \left[\cdots (\cdots) \cdots \right] \cdots \right\} .$$

Example. Compare:

$$\left\{ \int_a^b [f(x)]^p \, dx \right\}^{1/p} \qquad \left[\int_a^b \{f(x)\}^p \, dx \right]^{1/p}$$

From a conventional standpoint, the first one is better.

(j) Don't end a sentence with a symbol and start the next one with a symbol.

Example. Avoid this:

... where $s = 2\pi r$. T will denote tension ...

and especially this

... where $s = 2\pi r$. r is the radius ...

(k) Use words in place of logic symbols unless the use of logic symbols is standard in your area.

The logic symbols

symbol	meaning
$\forall$	for all, for every
$\exists$	there exists, there is
$:$	such that
$\therefore$	therefore, hence (conclusion indicator)
$\because$	because, since (premise indicator)
$\Rightarrow$	implies
$\Leftrightarrow$	if and only if
$\wedge$	logical and
$\vee$	logical or

some of which will be introduced in Section 7.3, are useful for thinking purposes. Unless you're writing specifically about mathematical logic, however, it's better to use words in formal engineering writing.

> **Example.** Compare for readability by an engineering audience:
>
> $\forall \varepsilon > 0, \exists N = N(\varepsilon): n > N \Rightarrow |a_n - a| < \varepsilon.$
>
> ✓ For every positive number ε, there is a number N (dependent on ε) such that $|a_n - a| < \varepsilon$ whenever $n > N$.

The first version might be appropriate for an audience of mathematicians, computer scientists, or even engineers in certain disciplines, but the second version is probably safer.

(1) For mathematically oriented text, consider using LATEX instead of a conventional word processor.

LATEX is a freely available macro package for the freely available mathematical typesetting program TEX. The present book was composed and typeset using LATEX, and many journals (mathematical, scientific, and technical) and book publishers now promote its use. Many university programs have policies stating that theses and dissertations must be done using LATEX.

LATEX does a particularly nice job typesetting mathematics. There is a learning curve involved, but we'll provide a one-page quickstart.

First, LATEX is not a WYSIWYG (*what you see is what you get*) system.[3] The user must prepare an input file, consisting of ordinary text along with some special LATEX commands, submit the file to the typesetting program, and then examine the typeset output. The typeset output cannot be directly changed on your computer screen. To change it you must change your input file, save the new version, run LATEX again, and examine the output again.

Here's a starter input file:

```
\documentclass{article}
\begin{document}
Hello world!
\end{document}
```

When run through LATEX, this file would result in a (mostly blank) page of typeset output that has "Hello world!" on it. Typesetting mathematics takes additional commands. For example, the structure

---

[3]Some add-on programs are available to make LATEX more friendly to the WYSIWYG enthusiast, but they are not part of LATEX itself.

```
\[
\frac{\alpha \beta \gamma}{\epsilon} = \int_a^b f(x) \, dx.
\]
```

produces the output

$$\frac{\alpha\beta\gamma}{\epsilon} = \int_a^b f(x)\, dx.$$

This, of course, represents just a few LaTeX math-mode commands. Here's a tiny sample of some others:

command	result	command	result
\delta	δ	\Delta	Δ
\pi	π	\Pi	Π
\cdots	$\cdots$	\pm	$\pm$
\rightarrow	$\rightarrow$	\odot	$\odot$
\le	$\le$	\subseteq	$\subseteq$
\approx	$\approx$	\hookrightarrow	$\hookrightarrow$
\nabla	∇	\partial	∂
\forall	$\forall$	\exists	$\exists$
\sum	$\sum$	\infty	∞
\oint	$\oint$	\Leftrightarrow	$\Leftrightarrow$
\ln x	$\ln x$	\cos x	$\cos x$
\in	$\in$	\sqrt{2}	$\sqrt{2}$
a_k	a_k	a^k	a^k

The page layout of a LaTeX document can be set by a *style file*: a file containing extra formatting commands. Many book publishers and journals now supply their own style files to authors. This allows the author to concentrate on writing rather than formatting.

A good way to learn more about LaTeX would be to install the program on your computer and play with it, using some of the many available explanations on websites and in books as guides. Typeset a page of text and mathematics, and then try getting the same look from your trusty old word processor. You'll see why the reader just might thank you for pulling yourself up the LaTeX learning curve.

Fortunately, certain commercial editors allow you to typeset equations in LaTeX and then port them directly to your word processor. Thus, your newly acquired typesetting skills may give some welcome flexibility in how you choose to write a document containing equations.

7.3 Writing Math Well

Writing math well means communicating ideas accurately and with minimal chance of misunderstanding. Formal mathematical logic has been developed to help you construct arguments leading to proofs, but the notation must be used properly and the principles understood. In this section we provide the tools needed to communicate your ideas effectively through mathematics.

Definitions

Definitions, along with primitive assumptions called *axioms* or *postulates*, form the starting points for logical argument. The subject of axioms is a sticky one best left to books on the foundations of mathematics.[4] But all engineers should know how to write acceptable definitions and recognize them (along with unacceptable ones) in the writing of others. A valid definition has two parts: (1) the term to be defined, and (2) the terms that do the defining.

> **Example.** The following is a definition.
>
> A *zigzog* is a hoontoon having sleebarbs at least two inches long.
>
> Here "zigzog" is the term to be defined, and the phrase "a hoontoon having sleebarbs at least two inches long" does the job of defining it.

A couple of comments:

1. The understanding of a definition may hinge on an understanding of other definitions. If we don't know what a *hoontoon* or a *sleebarb* or an *inch* is, we still won't know what a *zigzog* is. But that's *not* a fault of the definition above, which nicely specifies *zigzog* in terms of those other concepts!

2. A definition such as this one must be *simply accepted or not*. It is not to be proved or disproved; assuming the term *zigzog* wasn't defined previously (and is not in the common lexicon), an author is absolutely free to define it as shown. We do not argue as to the truth or falsity of definitions. We can *only* argue over whether they are *useful* or not. Depending on everyone's purposes (writer and audience), this may be a really bad way of defining *zigzog*. Or it may be a superb one. At this point, it's pretty hard to tell. But from a logical standpoint, there's nothing wrong with it as a definition.

---

[4]For a glimpse of that stickiness, peruse the first chapter of *Approximating Perfection: A Mathematician's Journey Into the World of Mechanics*, by L.P. Lebedev and M.J. Cloud, Princeton University Press, 2004.

Example. The following is a definition:

A *square* is a planar, four-sided figure.

Here *square* is the defined term and *planar, four-sided figure* is the defining phrase. Note, however, that the term *square* is in common use, and this definition differs from the conventional one. A *square* — as practically everyone in a technical audience will understand the term — must be a *rectangle and* all four of its sides must be equal in length. From that standpoint, we do not have a good definition in this case.

The example of the square points out something about the defining phrase of a good definition: it must in some way specify the *genus* and *species* (or *differentia*) of the term defined. That is, the defining phrase should somehow tell us the smallest *class* of objects to which the defined term belongs *and* specify something that contrasts it from all other elements of that class. The class *planar, four-sided figure* contains rectangles, parallelograms, trapezoids, and more general oblique figures, as well as squares! The definition specified the genus but not the species for the universally-accepted term *square*. Here are two better definitions of *square*:

A *square* is a rectangle with all sides equal.

A *square* is a rhombus with right angles.

Of course, the last definition requires the understanding of *rhombus*.

Example. Here are some pretty good, conventional definitions:

A *mallet* is a tool for striking other objects.

By the term **resistor**, we understand any two-terminal electrical device for which the terminal voltage and current are directly proportional with a positive constant of proportionality.

Let F be the magnitude of the total force acting on lever A.

Note our use of *italic font* or **bold font** with the defined term. This practice may not be necessary, but it is often desirable as a visual cue for the reader. Sometimes mathematical writers use embellished equals signs to indicate definitions:

$$x^2 := x \cdot x \,, \qquad x^2 \equiv x \cdot x \,, \qquad x^2 \triangleq x \cdot x \,,$$

and so on.

Is it possible to define *every* term you use? Clearly not: an attempt to do so would have you struggling to define words like *the* and *it*. But you should define any *crucial* term, *especially* if you use it in an unconventional sense.

Example. This author isn't expecting an audience of mind readers.

> Although a real number x is typically termed *positive* in the math literature if $x > 0$, in this monograph we will relax the latter condition to $x \geq 0$ (that is, we will use the term *positive* in cases where more elementary books would use the term *nonnegative*).

Imagine what confusion it might cause if he left out *that* little detail.

A typical textbook on logic will contain many pages on the fine points of definitions: their history, how they are classified, etc. The idea of recursive definition is treated in Exercise 7.29.

Logical Implication

As in Chapter 4 (recall p. 38) we denote statements or *propositions* by capital letters such as P, Q, R, and S. The *negation* of a statement P will now be denoted by $\sim P$. To simplify matters, we will adopt the usual convention in mathematics that a statement written down all by itself is taken to be true. So when we assert P, we mean that "P is true."

We also introduce the logical connectives *and* and *or*. The *conjunction* of two statements P and Q is the statement "P and Q". It is true if and only if both P and Q are true. The *disjunction* of two statements P and Q is the statement "P or Q". It is false if and only if both P and Q are false.

A statement of the form

$$\text{If } R, \text{ then } S \tag{A}$$

is called a *logical implication* in which R is the *hypothesis* (or *antecedent*) and S is the *conclusion* (or *consequent*). It can be expressed in various forms:

$$\text{If } R, \text{ then } S \quad \begin{array}{c} \text{means the same thing as} \\ \text{any of the statements} \end{array} \quad \left\{ \begin{array}{l} S, \text{ if } R \\ R \text{ only if } S \\ S \text{ whenever } R \\ R \text{ implies } S \\ S \text{ is implied by } R \\ R \text{ is sufficient for } S \\ S \text{ is necessary for } R \end{array} \right.$$

This list alone will enable you to safely rephrase a lot of mathematical writing.

Converse and Logical Equivalence

In general, given (A) we *cannot* turn around and assert that

> If S, then R.   (B)

The *converse* (B) of the implication (A) may or may not hold, depending on the natures of R and S. If (A) and (B) both hold, we write

> R if and only if S

and call the statements R and S *logically equivalent*. In English, this can be expressed in many different forms:

$$
R \text{ if and only if } S \quad
\begin{matrix} \text{means the} \\ \text{same thing} \\ \text{as any of} \end{matrix}
\left\{
\begin{matrix}
\text{If } R \text{ then } S \text{ } and \text{ if } S \text{ then } R \\
R \text{ implies } S, \text{ and conversely} \\
R \text{ and } S \text{ are logically equivalent} \\
R \text{ when and only when } S \\
R \text{ is necessary and sufficient for } S
\end{matrix}
\right.
$$

Example. Let's consider the following logical implication:

> If $x > 2$, then x is positive.

Here the hypothesis is $x > 2$, and the conclusion is x *is positive*. This implication is correct. Next consider the converse of the implication:

> If x is positive, then $x > 2$.

This is obviously incorrect, since the number 1 is positive, but it is not greater than 2. Thus, we would be incorrect to say

> x is positive if and only if $x > 2$.

Some Practice with Mathematical Arguments

So far, we have only given examples where P, Q, R, S are absolute statements; that is, each one was always true or false. In mathematics, of course, many statements involve variables, e.g.,

> $x - 1$ is a positive number.

Such a statement might be denoted by $P(x)$, and in order to get an absolute statement from it we would have to pick a certain value for x. Provided we consider only real numbers x, we see that $P(x)$ is true for all $x > 1$ and false for all $x \leq 1$. In contrast, the statement $Q(x)$ given by

x^2 is a nonnegative number

is true for any real value of x.

Example. Let's look at a simple but typical mathematical argument:

> The inequality $x^2 \geq 2x - 1$ holds for any real number x. To see this, we start with the known fact that $y^2 \geq 0$ for any real number y, and write y as $x-1$ to get $(x-1)^2 \geq 0$. Expanding the left-hand side and transposing two of the terms to the right-hand side, we establish the assertion.

In order to analyze this argument, we might display it schematically as

line	statement	justification
1	P: $y^2 \geq 0$ for any real y	known fact about the real numbers
2	Q: $y = x - 1$ for some x	definition of x, usual substitution idea
3	R: $(x-1)^2 \geq 0$	implied by lines 1 and 2
4	S: $x^2 \geq 2x - 1$	implied by 3 and a bit of algebra

So we start with two statements P and Q. We know that P holds by what mathematicians tell us about the real number system, and that it's fair game mathematically for us to introduce statement Q. The *conjunction* statement "*P and Q*" implies statement R (or, for example, is sufficient for R — remember all the ways we could say this in English and still be correct). After that, the laws of algebra taken together with statement R imply statement S, which is the desired conclusion of the argument.

Example. Let's analyze another simple argument:

> The inequality $x \geq 2 - 1/x$ holds for any real number x. To see this, we start with the known fact that $y^2 \geq 0$ for any real number y, and write y as $x - 1$ to get $(x-1)^2 \geq 0$. Expanding the left-hand side, transposing two of the terms to the right-hand side, and dividing through by x, we establish the assertion.

But something is wrong here because if we put $x = -1$ into the stated inequality $x \geq 2 - 1/x$, we get $-1 \geq 3$. Again, breaking the argument down schematically can be helpful.

line	statement	justification
1	P: $y^2 \geq 0$ for any real y	known fact about the real numbers
2	Q: $y = x - 1$ for some x	definition of x, usual substitution idea
3	R: $(x-1)^2 \geq 0$	implied by lines 1 and 2
4	S: $x^2 \geq 2x - 1$	implied by 3 and a bit of algebra
5	T: $x \geq 2 - 1/x$	implied by 4 *only* for $x > 0$ ... oops!

In algebra, we learned that when we divide both sides of an inequality by a negative number, we must reverse the inequality sign! Furthermore, it's not permissible to divide both sides of a given inequality by zero. So line 5 holds, as stated in the table, *only* for positive values of x. In short, we do not have "S implies T" and the chain of reasoning is broken at that point. The given argument was invalid because one of its implications was false. This is the blunder that Alexis made on p. 110.

Logical Quantifiers

Phrases such as "for all" and "there exists" appear throughout mathematical writing — even the mathematical writing done by engineers! They are known as *logical quantifiers*. Let's display some basic information about them in a table:

name	symbol	common English versions
universal quantifier	$\forall$	for all, for every, for each
existential quantifier	$\exists$	there exists, there is, there exist

The words "such that" often accompany the existential quantifier. So instead of writing

For every real number x, there is a real number y such that $x < y$

(obviously a true statement) we could write a bit more symbolically

$$\forall x \in \mathbb{R}, \exists y \in \mathbb{R} \text{ such that } x < y \tag{C}$$

where $\mathbb{R}$ is the set of real numbers and $x \in \mathbb{R}$ means x is an element of $\mathbb{R}$ (the usual notation of elementary set theory). Observe what happens if we reverse the order of the two quantifiers in (C):

$$\exists y \in \mathbb{R} \text{ such that } \forall x \in \mathbb{R}, \quad x < y. \tag{D}$$

This statement is false — can you display a real number y that is greater than every real number x? In other words, is there a greatest real number? We know there isn't; we can always add 1.78 to any real number and get a greater one. So the order of logical quantifiers can be crucial in your writing!

Negating a Quantified Statement

You should know how to correctly negate a quantified statement such as (D). In other words, you should be able to rephrase the true statement

It is *false* that there is a y such that for all x we have $x < y$ (E)

correctly in English. In more symbolic form, we have

$\sim [\exists y$ such that $\forall x, x < y]$.

The procedure is as follows. We move the negation sign ($\sim$) to the right; each time we pass a quantifier, we change it to the other type of quantifier:

$\forall y, \exists x$ such that $\sim [x < y]$.

Now we simply negate the final statement to get

$\forall y, \exists x$ such that $x \geq y$.

Hence we could rephrase (E) as

For each y, there is an x such that $x \geq y$.

Much easier to read!

Patterns of Mathematical Proof

Engineers are not mathematicians, but they must still be able to incorporate proper mathematical arguments into their work. Although they may not explicitly label their arguments as "proofs," these arguments must essentially *be* proofs in order to be valid. (Some engineers — notably in areas such as communication theory and control theory — *do* label their arguments as proofs.)

The basic activity of mathematical proof is that we take a collection of definitions, accepted axioms, and already established propositions, and, by valid patterns of inference, use these to establish a conclusion.

Direct Proof

The most common and desirable form of proof is the *direct proof*. Typically we are trying to establish statements of the form "P implies Q," and here the proof is laid out as follows.

> Assume that P holds. Then ..., and hence Q must also hold. Therefore P implies Q.

Of course, the part indicated by the three dots is where all the work lies.

Example. Suppose we have to show that if k is an even integer, then so is k^2. Here is a direct proof:

> Assume k is even. This means there is an integer n such that $k = 2n$. Then $k^2 = (2n)^2 = 2(2n^2) = 2p$, where p is also an integer. So k^2 can be written as $2p$, and hence it is even by definition. Therefore k^2 is even whenever k is even.

Note how it takes the general form exhibited in the box above.

Example. Here is a derivation of the voltage division rule from linear circuit theory. The rule states that if a voltage having value V_t is placed across a series combination of N resistors $R_1, \ldots, R_N$, then the voltage V_n across the nth resistor R_n is given by

$$V_n = V_t \left(\frac{R_n}{R_1 + \cdots + R_N} \right) \qquad (n = 1, \ldots, N) .$$

Although most engineers wouldn't call this passage a "proof," please pay attention to its general structure as an argument (we won't distract you with a circuit diagram).

> Suppose $R_1, \ldots, R_N$ are connected in series and that V_t is placed across the combination. By definition of the term series elements, the resistors all carry a common current I. Furthermore, since the equivalent resistance of the series combination is known to be $R_e = R_1 + \cdots + R_N$, we have
>
> $$I = \frac{V_t}{R_e} = \frac{V_t}{R_1 + \cdots + R_N} .$$
>
> By Ohm's law applied to the nth resistor R_n then, we obtain
>
> $$V_n = I R_n = \frac{V_t R_n}{R_1 + \cdots + R_N} ,$$
>
> which is the voltage division rule.

Note that we didn't explicitly assert "therefore P implies Q" at the end. Engineers often omit that formality (as do many mathematicians). Nonetheless, the argument form employed was that of direct proof.

Example. Let's stay with the voltage division rule momentarily, illustrating what *not* to do. The following passage is *invalid* as an argument form.

In order to prove "*P* implies *Q*," we *cannot* simply assume that *Q* is true and work backward to get *P* or some miscellaneous true statement. Although doing so can give us some insight that could be used to construct a valid argument, the final argument cannot be presented in such a way. That's not how this game is played. Ok, here we go:

> The relationship
>
> $$V_n = V_t \left(\frac{R_n}{R_1 + \cdots + R_N} \right)$$
>
> must be true because if we replace $V_t/(R_1 + \cdots + R_N)$ by I where I is the common current, we get $V_n = IR_n$. This is just Ohm's law for the nth resistor R_n, which we know to be true. Voltage division therefore holds.

Ouch! This "argument" contains some correct formulas and cites a correct definition ("series elements carry common currents") and a valid physical law (Ohm's law), but it is utterly wrong as an argument form. Why? Again, *our ability to reason from an assumption to a true conclusion does not establish the assumption.* Let's take a much simpler example. We'll invalidly "prove" that $-1 = 1$.

> The relationship $-1 = 1$ must be true, because if we square both sides we get $1 = 1$. This is obviously true, and we are done.

Double ouch.

Proof by Contrapositive

Given an implication

> If R, then S.　　　　　　　　　　　　　　　　　　　　　　(F)

we can form its *contrapositive*

> If $\sim S$, then $\sim R$.　　　　　　　　　　　　　　　　　　　(G)

These statements are logically equivalent (one is true when and only when the other one is), but in some cases it is easier to prove the contrapositive. Every mathematician knows this and so should every engineer.

Example. Suppose we want to prove that if the square of an integer is odd, then so is the integer. Symbolically,

$$\underbrace{k^2 \text{ is odd}}_{R} \text{ implies } \underbrace{k \text{ is odd}}_{S} \, .$$

Since an integer is either odd or even but not both, the contrapositive looks like

$$\underbrace{k \text{ is even}}_{\sim S} \text{ implies } \underbrace{k^2 \text{ is even}}_{\sim R} .$$

We could proceed to prove this latter statement as in Example 7.3.

Example. Here's the argument typically used to show that the electric field intensity at points interior to a conducting body in electrostatic equilibrium is zero. Before giving the argument, let's review some basic physics and get at the underlying idea. The conductor has free electrons that are available to move under the influence of an electric field of magnitude E. In fact, the force on an electron having charge e would have magnitude $F = eE$, and Newton's second law would predict an acceleration $a = F/m$ where m is the mass of the free electron. Acceleration means we don't have static equilibrium. Ok, here we go:

> If a conducting body is in electrostatic equilibrium, then at any point in its interior the electric field must vanish. Indeed, take a location P in the body and assume $E \neq 0$ there. Then an electron at P will experience a force eE and hence an acceleration. Since one of its constituent charges is accelerating, the body is *not* in electrostatic equilibrium. This establishes the claim.

Did you get that? Here it is again with a few explanatory notes:

> If a conducting body is in electrostatic equilibrium, then at any point in its interior the electric field must vanish (this is "R implies S"). Indeed, take a location P in the body and assume $E \neq 0$ there (this is $\sim S$). Then an electron at P will experience a force eE and hence an acceleration. Since one of its constituent charges is accelerating, the body is *not* in electrostatic equilibrium (we reached $\sim R$). This establishes the claim (we proved "$\sim S$ implies $\sim R$," so we can say that we proved "R implies S").

Sometimes authors will signal their readers with statements like

We proceed by contrapositive.

Let us prove the contrapositive instead.

Sometimes they won't, but that's a matter of English style rather than logical form. A general scheme for proof by contrapositive might look like this:

> In order to prove that P implies Q, we prove the contrapositive. Suppose $\sim Q$ holds. Then ..., and hence $\sim P$ must also hold. Therefore P implies Q.

The Inverse of an Implication

Note that the statement

If $\sim R$, then $\sim S$ (H)

which is called the *inverse* of (F), is *not* logically equivalent to (F). We cannot prove (F) by proving (H) instead. Let's summarize a few implication-related terms in a table:

name	statement	equivalent to "R implies S"
given implication	R implies S	yes, obviously
converse	S implies R	not in general
inverse	$\sim R$ implies $\sim S$	not in general
contrapositive	$\sim S$ implies $\sim R$	yes

In ordinary discourse, the inverse is often falsely labeled as the converse.

Example. Here *conversely* should be *inversely*:

> If the system operates correctly, the output power will be 2 Watts. Conversely, if it fails to operate correctly, the output power will be a different value.

Other Proof Methods

The arsenal of a mathematician is stocked with proof methods. We briefly indicate a couple more.

Proof by Contradiction

To prove

> R implies S,

we can suppose that R is true and S is false, and proceed to derive a *contradiction*. The contradiction can take any of the forms

1. $[T$ and $\sim T]$ (where T is any statement whatsoever),

2. $\sim R$ (since R was assumed true), or

3. S (since S was assumed false).

A proof done by contradiction can seem a little mysterious to a reader, which is why direct proofs are better when feasible. The day may arrive, however, when contradiction presents itself as the quickest route to a sound argument. A classic example of proof by contradiction is the proof that $\sqrt{2}$ is an irrational number. It starts by assuming that $\sqrt{2}$ is rational (which means it can be written in the form m/n where m and n are both integers) and proceeds to obtain a contradiction.

> Suppose there were a rational number $x = m/n$ such that $x^2 = 2$. Without loss of generality, we can assume m and n have no common factor (other than ± 1). Indeed, such a factor, if it existed, could be canceled and we could write $x = m/n$ where the new m and n values were mutually prime. Our supposition implies $m^2/n^2 = 2$, hence $m^2 = 2n^2$, from which we see that m^2 is even. But if m^2 is even, then so is m. Thus $m = 2p$, for some integer p, and $4p^2 = 2n^2$ so that $n^2 = 2p^2$. Therefore n^2 is even, hence so is n. Since m and m are both even, they each have two as a factor. This contradicts the hypothesis that m and n have no common factor. The proof by contradiction is finished.

Mathematical Induction

The formula

$$\sum_{k=1}^{n} k = \frac{n(n+1)}{2} \tag{3}$$

holds for all integers $n \geq 1$. Suppose we had to prove this. It certainly wouldn't suffice to try a few values of n and walk away satisfied. Equation (3) is an example of a mathematical proposition that we could call $P(n)$. We can prove $P(n)$ in two steps:

1. Show that $P(1)$ holds. That is, show that $P(n)$ is true for $n = 1$.

2. Show that $P(n)$ implies $P(n+1)$ for all $n \geq 1$.

Then we could say that

    $P(1)$ holds, and
    $P(1)$ implies $P(2)$ so $P(2)$ holds, and
    $P(2)$ implies $P(3)$ so $P(3)$ holds, and so on.

Therefore $P(n)$ holds for *every* $n \geq 1$. Step 1 is the *verification step* of mathematical induction, and step 2 is the *induction step*.

Let's try this with (3). First, $P(1)$ looks like $1 = 1$. Second, we *assume* $P(n)$ holds for some particular (but unspecified) integer $n \geq 1$:

$$\sum_{k=1}^{n} k = \frac{n(n+1)}{2} \ .$$

Adding $n+1$ to both sides, we obtain

$$\sum_{k=1}^{n+1} k = \frac{n(n+1)}{2} + (n+1) = \frac{(n+1)(n+2)}{2}$$

which is $P(n+1)$. We may now assert that $P(n)$ is true for all $n \geq 1$ by mathematical induction.

Existence and Uniqueness

Before leaving the subject of proof patterns, we should mention a couple of mathematical issues that frequently arise — even for engineers. They are

1. *existence* — does a problem have a solution?

2. *uniqueness* — if a problem has a solution, could it have more than one solution?

At first glance these may seem unexciting to an engineer, but consider the following:

1. Would you want to spend your entire career searching (with the help of a computer, say) for a solution to a problem if someone could prove rigorously that *there is no solution*? This is the issue of existence.

2. Would you want to find a solution to a problem but continue searching for another, better solution (however you might define "better" at the time) if someone could prove rigorously that *there is only one solution*? This is the issue of uniqueness.

So information regarding existence and uniqueness — the daily fare of many working mathematicians — can be crucially helpful to even the most practical-minded engineer. You may never have the occasion to establish either of these solution properties for yourself, but then again you may. Or you may see them addressed in the work of others. Engineers tend to be a curious lot, and we assume you're that way as well, so we offer a few more comments. How would someone prove existence of solution to a problem? He or she might

1. construct the solution and exhibit it, or

2. assume there is no solution and use that assumption to derive a contradiction,

but there are other clever ways as well. A common way of proving uniqueness is to assume the existence of two solutions, say x_1 and x_2, and then show that $x_1 = x_2$. Note that this (i.e., the act of assuming that solutions exist) would *not* establish existence, however, so existence and uniqueness are very different (but complementary) animals.

Uniqueness raises a point about the English language, regarding the articles *the* and *a*. By writing *the solution* instead of *a solution*, you could tacitly imply uniqueness where such has not yet been demonstrated and perhaps cannot be safely assumed. So be aware of that.

Example. Contrast the passage

> The solution to the governing equation, $x^2 = 1$, is $x = 1$.

with the passage

> The governing equation $x^2 = 1$ has two real solutions $x = \pm 1$. We choose the positive solution $x = 1$ as the one of physical interest in our problem.

Which one was more careful? Can you really just talk about *the solution* to a quadratic equation such as $x^2 = 1$?

Careless assumptions regarding existence can cause real trouble.

Example. Consider this argument (known as *Perron's paradox*):

> Suppose there exists a greatest positive integer N. Since N^2 is also a positive integer we must have $N^2 \le N$, from which it follows that $N = 1$.

So by a baseless existence-type assumption, we managed to conclude that there are no positive integers bigger than 1. Knowing what we know about the integers, we wouldn't be fooled by this. *However, if we knew nothing about the integers we might accept the bogus result and try to base our life's work on it.* Such things have happened in the mathematical sciences.

An Existence/Uniqueness Quantifier

Statements regarding existence and uniqueness properties are sometimes combined into the quantifier symbol "$\exists!$". So

$\exists! \, x$ such that $P(x)$.

is short for

There is a *unique* x such that $P(x)$ holds.

Example.

$\exists! x$ such that $x + 2 = 0$.

The unique value in question is, of course, $x = -2$.

7.4 The Value of Abstraction

We know that it's not permissible to add apples to oranges, as these objects are dissimilar. Nonetheless, something crucially important is shared by three apples and three oranges: the concept of *three*. Probably few engineers are totally against the notion of *three* simply because it is abstract. Yet engineers (and engineering students) do vary widely in their views on abstraction. Some shun abstraction and do everything possible to confine their thinking to the realm of concrete objects. Others see abstraction as a fruitful process and are always on the lookout for opportunities to use the power of generality to their advantage. In this final section of the book, we offer some suggestions on generality and abstraction in engineering, especially as these pertain to engineering writing.

Abstraction is the process of identifying general characteristics of specific objects, and then considering these characteristics apart from the objects themselves. Mathematics, of course, is largely concerned with abstraction; we see this in the most elementary notions of *number* and *set*, then up through algebra, calculus, and beyond. Understanding that there are time rates of change and space rates of change, mathematicians take the leap to the general notion of *rate of change* and hence to the first derivative as a calculation tool. Without abstraction, mathematics as we know it would be impossible. The same is true of logic. In Chapter 4, we used the idea of *argument form* to look for invalid arguments. An argument can be invalid simply because of its form, regardless of the concrete meanings of the statements it contains. The argument form

All P is M.
Some S is M. $\qquad\qquad\qquad\qquad\qquad$ × invalid
Therefore, some S is P.

is one such case. It represents an abstraction taken from infinitely many possible arguments along the lines of

All poodles are furry.
Some cats are furry.   × invalid
Therefore, some cats are poodles.

While it's sometimes easier to spot an invalid argument in such a concrete form, abstraction gives us the ability to cover infinitely many situations as specific instances of what is essentially a single case.

Many other sciences are deeply concerned with abstraction. What drives the quest for generality? First, greater generality means wider applicability and hence greater power and economy of effort. Newton's second law

$$f = ma \ ,$$

for example, neatly summarizes the results of innumerable concrete experiments. It certainly simplifies the presentation of classical mechanics. Second, generality can deepen existing knowledge and point to connections between seemingly unrelated areas. When the laws of electromagnetism were generalized to Maxwell's equations, for example, the connection between electromagnetics and optics was discovered. For many people (especially mathematically minded people), generality can also have significant aesthetic value. On the other hand, generality can come at a cost. One may have to learn (or invent) new ways of thinking in order to generalize a familiar structure, and the time and energy expenditure may not be warranted in a given set of circumstances.

What does all this mean to the engineer? On a very practical level, the more abstract notions and principles are the ones that provide the working engineer with a fundamentals base from which it is possible to cope with technological change. Physical and mathematical principles do change with new discoveries, but they tend to change more slowly than technological details do. An ability to think abstractly can also lend greater value and wider applicability to your work. Concrete specifics do not always the represent the best route to clarity.

Example. Tanya works at a company that manufactures electronic devices. She has been given the job of cataloging and posting to the web the large number of standard laboratory procedures for testing the devices. These procedures have such titles as "vibration" and "lifespan." At first the prospect of organizing over 100 procedures in a meaningful way seems overwhelming, but she quickly generates some ideas. The most straightforward seems to be to organize the procedures by the device being measured. But there are a multitude of devices and many are covered by more than one procedure. Instead she comes up with a categorization that is more abstract, involving the principles of operation of the devices rather than the devices themselves. For instance, categorizing "active devices" rather than transistors, op amps, SCRs, TRIACs, and tubes recognizes that these share many common measurements, such as thermal dissipation, current flow, and lifespan. This organization works quite well, since the users of the procedures naturally think along these lines of abstraction, and look for such commonalities.

Next, whenever it's time to present background theory in a formal, technical document, the engineer faces some decisions regarding what level of generality will be of most value to the reader.

Example. John, a mechanical engineering postdoc, plans to base a journal paper on his doctoral dissertation. Since his target journal is read by experts in fluid dynamics, John will try to present the necessary theoretical background with a fair amount of generality. He knows that the theory he developed could be useful to workers in related areas and he doesn't wish to inhibit the spread of information by particularizing his discussion too much. John also understands that if he fails to present his theory in the most general form he knows, then *someone else* may publish the theory in that form and claim it as his own.

Mathematicians and mathematically oriented engineers often value general theorem statements. Again, this may be for aesthetic reasons but it may also be due to the increase in power afforded by the more general statements. As an engineer writing for other engineers, you will have to use discretion. If generality detracts or will certainly not be useful, then by all means limit your discussion to a more concrete case.

Example. Mary is an electrical engineering professor writing a textbook on electric circuits for use at the junior level. Although Mary has an expert grasp of general network theory, she keeps her target audience in mind as she writes. There is no point in overwhelming undergraduate students with more theory than they can handle given their present level of academic maturity and knowledge of the field. Mary also understands that other instructors may not adopt her book unless it falls within accepted norms. So she aims for a level that she believes will represent the best compromise between generality and concreteness. The target reader should be able to use her book as a springboard to books at the next level of abstraction.

7.5   Chapter Recap

1. Although engineers are not mathematicians, they must be able to write mathematical arguments.

2. Like any other type of technical claim, a mathematical claim must be verifiable.

3. Great care is warranted with mathematical terminology.

4. Standard use of mathematical symbolism is recommended.

5. LaTeX is a freely available mathematical typesetting system.

6. A valid definition has two parts: (1) the term to be defined, and (2) the terms that do the defining.

7. A proposition is something that can be proved true or false. A definition is not a proposition; a definition can be clear and useful, but not true or false.

8. You should define any crucial term that is either used unconventionally or unfamiliar to the target reader.

9. An implication is a statement of the form *If P, then Q*. There are many ways to rephrase such statements in English.

10. An equivalence is a statement of the form *P if and only if Q*. Such a statement can be rephrased in many ways.

11. Along with a given implication, we have its converse, its inverse, and its contrapositive. Only the contrapositive is necessarily equivalent to the given implication.

12. The basic idea of direct proof is to start with definitions, standard assumptions, and already established propositions, and work toward the desired conclusion. This is not the same as starting with the desired conclusion and working toward some other statement.

13. Besides direct proof, other proof schemes are available. These include proof by contrapositive and proof by contradiction.

14. Mathematical induction is like falling dominoes. You show that if one falls, then the next one has to fall. By showing that the first one falls, you show that they all fall.

15. The existential and universal quantifiers appear throughout mathematical writing, including engineering writing. It's important to understand the rules that govern their use.

16. It's easy to blunder and mislead when dealing with issues of mathematical existence and uniqueness. The fact that you found a solution — and call it *the* solution — doesn't mean it's the only one that exists. The fact that you can talk about a solution — and give it a name such as x_0 — doesn't mean it exists.

17. Generality and abstraction can increase the efficiency and effectiveness of engineering discourse. They have great value when used appropriately.

7.6 Exercises

7.1. What is the difference between a *circle* and a *circular disk*? Repeat for *sphere* and *spherical ball*. Repeat for *positive number* and *nonnegative number*.

7.2. Explain the following mathematical terms: *integer, real number, rational number, irrational number*.

7.3. Fix each sentence.

(a) Let us graph the expression $4x^2$.

(b) We shall consider the expression $x^2 = 5$.

(c) The identity $\sin x$ is often useful in electronics.

(d) It is a simple matter to solve the quadratic equation $ax^2 + bx + c$.

(e) The function $x^2 = 10$ can be solved by Newton's method.

(f) We shall denote this length by Ω.

(g) After simplification, we obtain $\ln \frac{a}{b}$.

(h) We have $f = m \odot a$ where $\odot$ denotes multiplication.

(i) We have $y^2 + 3y - 4 = 0 = y = 1$ or -4.

(j) The area of a circle is πa^2 where $a =$ the radius of the circle.

(k) Simplifying, we obtain $x = (3\sin\{y\})^2$.

(l) This is not the smallest possible value of p. p can be decreased through the use of another method.

(m) Since $d^2 + c^2 = e^2 \rightarrow e = d$ ($\because c = 0$).

(n) If $x > 0$, $x^2 > 0$.

(o) Squaring the expression xy, we obtain $x^2 y$.

(p) C is the boundary contour of S. C is therefore closed.

(q) A square of side length L has area $A = L \cdot L$.

(r) A square of side length L has area $A = L^2$.

(s) Convergence of the algorithm was seen after five iterations.

(t) The area of a circle is given by $\pi(r)^2$.

(u) The rectangle is L long and W wide.

(v) Let `machine_epsilon` be a small number.

7.4. Code the following equations in LaTeX:

(a) $H_n^{(1)}(x) = J_n(x) + jY_n(x)$

(b) $(1 - x^2)y''(x) - 2xy'(x) + \lambda y(x) = 0$

(c) $R_n(r) = A_n r^n + \dfrac{B_n}{r^{n+1}}$

(d) $P_{2n}(0) = \dfrac{(-1)^n (2n)!}{2^{2n} (n!)^2}$

(e) $r^2 \dfrac{d^2 R}{dr^2} + 2r \dfrac{dR}{dr} - n(n+1)R = 0$

(f) $u(x,t) = \displaystyle\sum_{n=1}^{\infty} E_n \sin nx\, e^{-n^2 kt}$

(g) $\displaystyle\int_0^{2\pi} \int_0^{\pi} Y_{l'm'}^*(\theta, \phi) Y_{lm}(\theta, \phi) \sin\theta\, d\theta\, d\phi = \delta_{l'l}\delta_{m'm}$

(h) $\displaystyle\int_0^1 \ln\left(\dfrac{1+x}{1-x}\right) \dfrac{dx}{x} = \dfrac{\pi^2}{4}$

(i) $\dfrac{1}{4\pi\epsilon} \displaystyle\int_S \dfrac{\rho_s(\mathbf{r}')\, dS'}{|\mathbf{r} - \mathbf{r}'|} = V_0$

(j) $A = \begin{pmatrix} 3 & 5 \\ -2 & -4 \end{pmatrix}$

7.5. Learn how to create an index entry in a LaTeX document.

7.6. Some engineering writers like to append long, explanatory subscripts to their mathematical symbols. Thus, instead of reading

> Recall that $f = ma$, where f is the net force acting on the body, m is the mass of the body, and a is the acceleration of the center of mass of the body.

we may have to read

> Recall that $f_{\text{net on body}} = m_{\text{body}} a_{\text{center of mass}}$.

Comment on this practice. Would you prefer to read much material written in this style?

7.7. Criticize the following definitions.

(a) A *bee* has a stinger on the tip of its abdomen.

(b) A *triangle* consists of three straight lines.

(c) A *force* is a push or a pull.

7.8. Write down your best definition of the term *engineering* (cf. Exercise 2.5). Repeat for a more specific term such as *electrical engineering, mechanical engineering,* or *civil engineering*.

7.9. A statement P is said to be *stronger* than a statement Q if P implies Q but it is false that Q implies P. We also say that Q is *weaker* than P. Which of the following statements is stronger?

R: x is a real number and $x < 1$.
S: x is a real number and $x < 2$.

7.10. Rephrase the implication

If $x > 1$, then $x^2 > 1$

a few different ways in English. Repeat for the equivalence

We have $|x| < 1$ if and only if $-1 < x < 1$.

7.11. Formulate the converse, inverse, and contrapositive of each statement:

(a) If Lisa is an engineer, she is smart.
(b) Lisa is smart only if she is an engineer.
(c) Lisa is smart if she is an engineer.

7.12. Given the statement "p implies q," find

(a) the contrapositive of the converse,
(b) the inverse of the converse,
(c) the contrapositive of the inverse,
(d) the converse of the inverse,
(e) the converse of the contrapositive,
(f) the inverse of the contrapositive.

7.13. Write a true statement whose converse is false. Write a true statement whose converse is also true.

7.14. Let x be a real number. True or false:

(a) If $x < 1$, then $x^2 < 1$.
(b) If $x^2 < 1$, then $x < 1$.
(c) We have $x^2 < 1$ if and only if $x < 1$.

7.15. Observe that a universally quantified statement can be written as an implication. For example, the statement

For every real number x, we have $x^2 \geq 0$.

can be rephrased as

If x is a real number, then $x^2 \geq 0$.

Generate an example of your own.

7.16. Negate each statement below:

(a) The function is real and has even symmetry.
(b) The solution is real but not rational.
(c) The solution is either real or complex.
(d) For every $\varepsilon > 0$, there exists $\delta > 0$ such that for all $x \in [a, b]$ we have $|f(x) - f(x_0)| < \varepsilon$ whenever $|x - x_0| < \delta$.
(e) For every $\varepsilon > 0$, there exists $\delta > 0$ such that for all $x \in [a, b]$, $|x - x_0| < \delta$ implies $|f(x) - f(x_0)| < \varepsilon$.

7.17. Find a short derivation in your favorite engineering textbook. Rewrite it in your own words, making sure to develop a valid argument form.

7.18. What is meant by the term *vicious circle*? Construct an example.

7.19. Prove by mathematical induction: $1 + 3 + 5 + \cdots + (2n - 1) = n^2$.

7.20. Consideration of the English article "a" shows why care in writing is necessary. Do the two statements

> We recommend the use of a toggle and safety mechanism.
> We recommend the use of a toggle and a safety mechanism.

communicate exactly the same thing? Why or why not?

7.21. Criticize the following mathematical argument.

If a, b, and x are positive numbers with $a < b$, then

$$\frac{a}{b} < \frac{a + x}{b + x} . \tag{4}$$

Indeed, starting with (4) we can cross-multiply and obtain

$$a(b + x) < b(a + x)$$

which simplifies to $a < b$ as assumed.

7.22. Is the following mathematical argument valid?

If a, b, and x are positive numbers with $a < b$, then

$$\frac{a}{b} < \frac{a + x}{b + x} . \tag{5}$$

Indeed, suppose $a < b$. Multiplying through by x, we obtain $ax < bx$. Adding ab to both sides of this and factoring, we obtain

$$a(b + x) < b(a + x) .$$

Slight rearrangement yields (5), as desired.

7.23. Prove the *inequality of the means*

$$\sqrt{ab} \le \frac{a + b}{2}$$

for positive real numbers a and b.

7.24. Here's a classic argument. What's wrong with it?

If $x = 1$, then $x = 0$. Indeed, suppose $x = 1$. Multiplying both sides by x, we get $x^2 = x$. Subtracting 1 from both sides, we get $x^2 - 1 = x - 1$. Now, factoring the left-hand side as $(x + 1)(x - 1)$ and canceling the factor $x - 1$ from both sides, we get $x + 1 = 1$. This shows that $x = 0$ as desired.

7.25. What's wrong with this argument?

We have $1 = -1$. Indeed, let us start with $(-1)(-1) = (-1)^2$. Taking the square root of both sides, we obtain

$$\sqrt{(-1)(-1)} = (\sqrt{-1})^2$$

which yields $\sqrt{1} = i^2$ where i is the elementary imaginary number. Since $i^2 = -1$, the proof is complete.

7.26. What's wrong with this argument?

We have $1 = -1$. Indeed, let us start with $(-1)^2 = 1$. Taking the natural log of both sides, we obtain $2\ln(-1) = 0$ so that $\ln(-1) = 0$. We complete the proof by raising e to the power of both sides.

7.27. What's wrong with this argument?

The equation $x - 3 = 2$ has two solutions: $x = 1$ and $x = 5$. It is immediately obvious that $x = 5$ satisfies the equation. Substituting $x = 1$ and squaring both sides, we get $(1 - 3)^2 = 2^2$ and hence $4 = 4$, which is also true. Both solutions are thereby verified.

7.28. Let $f(x)$ be a real-valued function of a real variable x, and suppose $f(x)$ is differentiable at $x = x_0$. Is the condition $f'(x_0) = 0$ a necessary one for x_0 to be an extreme point of $f(x)$? Is it a sufficient condition?

7.29. Functions defined on the natural numbers are sometimes defined *recursively*. A recursive definition has two components: (1) a *base case*, and (2) an *inductive rule*. For a function $f(n)$, the base case may be specification of $f(0)$ and the inductive rule a specification of $f(n+1)$ in terms of $f(n)$. The elementary factorial function $n!$, for instance, can be defined by the two conditions $0! = 1$ and $(n+1)! = (n+1) \cdot n!$. More generally, we need a way to get the recursion started, and a way to get each function value from its predecessor(s). Use the pair of conditions $g(n) = g(n-1) + g(n-2)$ and $g(1) = g(2) = 1$ to compute the first few *Fibonacci numbers* $g(n)$.

7.30. Recall that the empty set is the set with no elements. Any statement we make about the empty set is true. Explain, giving at least one example.

Further Reading

Technical Writing

We mentioned that you can find much more comprehensive books on technical writing. Some more recent titles include

Writing for Science and Engineering (Second Edition), by Heather Silyn-Roberts. Elsevier, 2013.

Scientific Papers and Presentations (Third Edition), by Martha Davis. Elsevier, 2013.

The Manual of Scientific Style: A Guide for Authors, Editors, and Researchers, edited by Harold Rabinowitz and Suzanne Vogel. Elsevier, 2009.

Chicago Guide to Communicating Science, by Scott L. Montgomery. University of Chicago Press, 2002.

Engineers' Guide to Technical Writing, by Kenneth G. Budinski. ASM International, 2001.

Science and Technical Writing: A Manual of Style, edited by Philip Rubens. Routledge, 2001.

English for Writing Research Papers, by Adrian Wallwork. Springer, 2011.

Older but insightfully written books include

Technical Writing: Principles, Strategies, and Readings, by Diana C. Reep. Allyn & Bacon, 1997.

Technical Writing, by John M. Lannon. HarperCollins, 1988.

Scientists Must Write: A Guide to Better Writing for Scientists, Engineers and Students, by Robert Barrass. Chapman & Hall/CRC Press, 1978.

Handbook of Better Technical Writing, by Dudley H. Rowland. Business Reports, Inc., 1962.

Technical Writing, by T.A. Rickard. Wiley, 1920.

A broader book on general technical communication (including oral presentation) is

Essential Communication Strategies for Scientists, Engineers, and Technology Professionals, by Herbert L. Hirsch. IEEE Press, 2003.

The books

A Handbook for Analytical Writing: Keys to Strategic Thinking, by William E. Winner. Morgan & Claypool, 2013.

Clear and Concise Communications for Scientists and Engineers, by James G. Speight. CRC Press, 2012.

have chapters on teamwork in multi-author projects. The book

Guidelines for Writing Effective Operating and Maintenance Procedures, Center for Chemical Process Safety, American Institute of Chemical Engineers, 1996.

includes a chapter on writing emergency procedures. Titles that deal with the ethics as well as the mechanics of technical writing include

Style and Ethics of Communication in Science and Engineering, by Jay D. Humphrey and Jeffrey W. Holmes. Morgan & Claypool, 2008.

Eloquent Science: A Practical Guide to Becoming a Better Writer, Speaker, and Atmospheric Scientist, by David M. Schultz. Springer, 2009.

The books

A Scientific Approach to Scientific Writing, by John Blackwell and Jan Martin. Springer, 2011.

From Research to Manuscript: A Guide to Scientific Writing, by Michael Jay Katz. Springer, 2009.

offer strategies for dealing with the comments of editors and peer reviewers. A nice book with an unconventional slant and some useful insights is

Scientific Writing 2.0: A Reader and Writer's Guide, by Jean-Luc Lebrun. World Scientific, 2011.

We like the way the book

Scientific Writing: Thinking in Words, by David Lindsay. CSIRO Publishing, 2010.

is organized into parts called *Thinking about Your Writing* and *Writing about Your Thinking*. The book

Writing for Science, by Robert Goldbort. Yale University Press, 2006.

contains discussions of laboratory notebooks, undergraduate reports, and dissertations. It also has an interesting section called "Measuring Scientific Readability."

Writing Mathematical Arguments

Books that focus on mathematical writing and proof techniques include

A Transition to Advanced Mathematics, by Douglas Smith, Maurice Eggen, and Richard St. Andre. Cengage Learning, 2010.

Introduction to Mathematical Structures and Proofs, by Larry J. Gerstein. Springer, 2012.

A First Course in Abstract Mathematics, by Ethan D. Bloch. Springer, 2011.

The Art of Proof: Basic Training for Deeper Mathematics, by Matthias Beck and Ross Geoghegan. Springer, 2010.

Reading, Writing, and Proving: A Closer Look at Mathematics, by Ulrich Daepp and Pamela Gorkin. Springer, 2011.

The Nuts and Bolts of Proofs, Fourth Edition: An Introduction to Mathematical Proofs, by Antonella Cupillari. Academic Press, 2012.

How to Read and Do Proofs: An Introduction to Mathematical Thought Process, by Daniel Solow. Wiley, 2009.

One way to hone reasoning skills is to read *sophisms*: arguments intended to deceive. For this we recommend

Lapses in Mathematical Reasoning, by V.M. Bradis, V.L. Minkovskii, and A.K. Kharcheva. Dover Publications, 1999.

Mathematical Fallacies and Paradoxes, by Bryan H. Bunch. Dover Publications, 1997.

Science Writing

The game of popular science writing is covered in

A Field Guide for Science Writers, edited by Deborah Blum, Mary Knudson, and Robin Marantz Henig. Oxford University Press, 2006.

Communicating Science: A Practical Guide, by Pierre Laszlo. Springer, 2006.

Ideas into Words: Mastering the Craft of Science Writing, by Elise Hancock. Johns Hopkins University Press, 2003.

Old Favorites

Several decades ago, one could sample good technical writing by opening almost any published engineering textbook. Unfortunately, we believe this is no longer the case; economic, educational, and professional conditions have changed, and the textbook literature now exhibits wild variation in expository quality. The same statement holds for journals, which have proliferated in the electronic age. Therefore we'd steer you toward older works for examples of what we like to see. For example, we like the books in the old McGraw-Hill and Wiley series on electrical engineering (our particular field). Two titles recently reissued by the IEEE Press are

Time-Harmonic Electromagnetic Fields, by Roger F. Harrington.

Antenna Theory and Design, by Robert S. Elliott.

Some of the really beautiful technical books are far older; you might peruse, for instance,

Hydrodynamics (Sixth Edition), by Sir Horace Lamb. Cambridge University Press, 1993.

A Course in Pure Mathematics (Tenth Edition), by G.H. Hardy. Cambridge University Press, 2008.

and ground yourself a bit in an even more old-fashioned style. One of our favorite examples of good clear mathematical writing is

Finite Dimensional Vector Spaces, by Paul Halmos. Springer, 1974.

Engineering Design

For discussions of the engineering design process, we recommend

Fundamental Concepts in Electrical and Computer Engineering with Practical Design Problems, by Reza Adhami, Peter M. Meenen, and Denis Hite. Universal Publishers, 2007.

Creative Engineering Design, by Brian S. Thompson. Okemos Press, 1996.

English Grammar, Style, and Vocabulary

The reader who is interested in English grammar and style could consult such books as

The Elements of Style, by William Strunk Jr.

Errors in English and Ways to Correct Them, by Harry Shaw. Collins Reference, 1993.

The Most Common Mistakes in English Usage, by Thomas E. Berry. McGraw Hill Professional, 1971.

Style: Toward Clarity and Grace, by Joseph M. Williams. University of Chicago Press, 1995.

Grammar for Journalists, by E.L. Callihan. Chilton Book Company, 1979.

We also recommend having an unabridged print dictionary such as

Random House Webster's Unabridged Dictionary. Random House Reference, 2005.

For some interesting historical background on English vocabulary, we recommend

English Vocabulary Elements by Keith Denning, Brett Kessler, and William R. Leben. Oxford University Press, 2007.

Words Words Words, by C.M. Matthews. Macmillan, 1980.

Logic and Critical Thinking

For clear introductions to the principles of formal logic, see

Introduction to Logic (14th Edition), by Irving M. Copi, Carl Cohen, and Kenneth McMahon. Pearson, 2010.

Logic for Mathematicians, by J. Barkley Rosser. Dover, 2008.

For less formal discussions under the heading of "critical thinking," see, e.g.,

How the Great Scientists Reasoned: The Scientific Method in Action, by Gary G. Tibbetts. Elsevier, 2013.

An Introduction to Critical Thinking and Creativity: Think More, Think Better, by Joe Y.F. Lau. Wiley, 2011.

Workbook for Arguments: A Complete Course in Critical Thinking, by David R. Morrow and Anthony Weston. Hackett Publishing Company, 2011.

Critical Inquiry: The Process of Argument, by Michael Boylan. Westview Press, 2009.

A classic book about heuristic reasoning in mathematics is

How to Solve It: A New Aspect of Mathematical Method, by George Polya. Princeton University Press, 2004.

Persuasive Writing

See, for example,

> *Persuasive Business Proposals: Writing to Win More Customers, Clients, and Contracts*, by Tom Sant. AMACOM Books, 2012.

LaTeX and TeX

Interested in LaTeX for mathematical typesetting? The standard manual was written by the inventor himself:

> *LaTeX: A Document Preparation System*, by Leslie Lamport. Addison-Wesley, 1994.

Also useful is

> *A Guide to LaTeX: Document Preparation for Beginners and Advanced Users (3rd Edition)*, by Helmut Kopka and Patrick W. Daly. Addison-Wesley, 1999.

The TeX system on which LaTeX is based was invented by the computer scientist Donald Knuth. His book is

> *The TeXBook*, by Donald E. Knuth. Addison-Wesley, 1984.

Online Reference Documents

The following useful reference documents were available online at the time of publication of the present book.

> *The International System of Units (SI)*, edited by Barry N. Taylor and Ambler Thompson. NIST Special Publication 330, National Institute of Standards and Technology, 2008.

> *Dictionary of Occupational Titles*, US Department of Labor.

Quick Reference

Valid Categorical Syllogisms[†]

All M is P.	No M is P.	All M is P.
All S is M.	All S is M.	Some S is M.
$\therefore$ All S is P.	$\therefore$ No S is P.	$\therefore$ Some S is P.
No M is P.	No P is M.	All P is M.
Some S is M.	All S is M.	No S is M.
$\therefore$ Some S is not P.	$\therefore$ No S is P.	$\therefore$ No S is P.
No P is M.	All P is M.	All P is M.
Some S is M.	Some S is not M.	No M is S.
$\therefore$ Some S is not P.	$\therefore$ Some S is not P.	$\therefore$ No S is P.
Some M is P.	All M is P.	Some P is M.
All M is S.	Some M is S.	All M is S.
$\therefore$ Some S is P.	$\therefore$ Some S is P.	$\therefore$ Some S is P.
Some M is not P.	No M is P.	No P is M.
All M is S.	Some M is S.	Some M is S.
$\therefore$ Some S is not P.	$\therefore$ Some S is not P.	$\therefore$ Some S is not P.
All M is P.	No M is P.	No P is M.
All M is S.	All M is S.	All M is S.
$\therefore$ Some S is P.	$\therefore$ Some S is not P.	$\therefore$ Some S is not P.
	All P is M.	
	All M is S.	
	$\therefore$ Some S is P.	

[†] Without further conditions stated, the syllogisms in the final two rows are not considered valid in modern logic where empty sets are permitted. Those in the sixth row only hold if the class M is nonempty (i.e., there is an element x such that x belongs to M). The lone argument in the seventh row only holds if P is nonempty.

Examples of *Invalid* Categorical Syllogisms

Caution: These are *not* forms of valid reasoning. They are collected here only to put the reader on alert (see Exercise 4.3).

All P is M.	All M is P.	All M is P.
Some S is M.	No M is S.	All M is S.
∴ Some S is P.	∴ No S is P.	∴ All S is P.
No M is P.	Some P is M.	Some M is not P.
All M is S.	Some S is M.	Some S is M.
∴ No S is P.	∴ Some S is P.	∴ Some S is P.
Some M is P.	No P is M.	Some M is not P.
Some S is not M.	No S is M.	No S is M.
∴ Some S is not P.	∴ No S is P.	∴ Some S is not P.
No M is P.	No M is P.	All M is P.
All S is M.	Some S is M.	All S is M.
∴ All S is P.	∴ Some S is P.	∴ No S is P.
All M is P.	Some M is P.	All M is P.
Some S is M.	All S is M.	Some S is M.
∴ Some S is not P.	∴ All S is P.	∴ All S is P.
No M is P.	Some M is not P.	Some M is not P.
Some S is M.	All S is M.	All M is S.
∴ No S is P.	∴ No S is P.	∴ No S is P.

Valid Arguments Involving Conditional Statements

Modus Ponens

$$P \text{ implies } Q$$
$$P$$
$$\therefore Q$$

Modus Tollens

$$P \text{ implies } Q$$
$$\sim Q$$
$$\therefore \sim P$$

Hypothetical Syllogism

$$P \text{ implies } Q$$
$$Q \text{ implies } R$$
$$\therefore P \text{ implies } R$$

Formal Fallacies Involving Conditional Statements

Fallacies are *not* forms of valid reasoning. Stay away from these patterns in your writing, speaking, and thinking.

Affirming the Consequent

$$P \text{ implies } Q$$
$$Q$$
$$\therefore P$$

Denying the Antecedent

$$P \text{ implies } Q$$
$$\sim P$$
$$\therefore Q$$

Some Informal Fallacies

Again, these are *not* forms of valid reasoning and must be avoided!

Ad Hominem. Arguing against the person by attacking or discrediting him, or alluding to his possible motives.

Fallacy of Accident. Trying to apply a rule to a case it wasn't intended to cover.

Straw Man. Distorting someone else's position and then attacking the distorted version.

Appeal to Ignorance. Saying, for example, that you can't imagine anything other than A causing B, hence A must have caused B.

Hasty Generalization. Trying to draw a conclusion about all members of a group from the characteristics of an insufficient sample.

Post Hoc Ergo Propter Hoc. Asserting that A must have caused B because A preceded B in time.

Cum Hoc Ergo Propter Hoc. Asserting that A must have caused B because A and B occurred simultaneously.

Fallacy of Composition. Erroneously attributing a trait possessed by all members of a class to the class itself.

Fallacy of Division. Erroneously attributing the traits of a class of objects to each of the separate objects.

Begging the Question. Using the conclusion you're trying to establish as one of your premises.

Weak Analogy. Trying to argue based on alleged similarity between two situations that, in reality, are not that similar.

False Dichotomy. Basing an argument on the premise that either A or B must hold, when in reality a third possibility C might hold.

Fallacy of Suppressed Evidence. Omitting counterinstances while drawing an inductive conclusion.

Fallacy of Equivocation. Using a word in two different ways in the same argument.

Fallacy of Amphiboly. Arguing based on a faulty interpretation of an ambiguous statement.

Appeal to the Crowd. Arguing that A must be true because most people believe it's true.

Fallacy of Opposition. Arguing that A must be false because your opponent believes it's true.

Appeal to Authority. Arguing that A must be true because experts believe it's true.

Affirming the Consequent. Arguing that a logical implication is true because its converse is true.

Denying the Antecedent. Arguing that a logical implication is true because its inverse is true.

Non-Sequitur. Arguing so loosely that your conclusion simply does not follow from your premises.

Checklists for Evaluating the English Content of a Formal Engineering Document

See Chapters 5 and 6 for expanded treatment of these items.

- ☐ Each paragraph accomplishes a task.
- ☐ Every declarative sentence has a subject and a predicate.
- ☐ Each sentence is of reasonable length.
- ☐ Important ideas are made grammatically important.

☐ The active voice and strong verbs are used for vigor.

☐ There is adequate variety in sentence structure and length.

☐ The subjunctive mode signals conditions contrary to fact.

☐ Every sentence is punctuated correctly.

☐ Proper capitalization is implemented throughout.

☐ The wording is generally concise and direct.

☐ Opportunities for parallel construction are exploited.

☐ All abbreviations are standard.

☐ All comparatives are used properly.

☐ Grammatical agreement in number is implemented throughout.

☐ Every pronoun refers clearly to its intended antecedent.

☐ Every modifier is placed near the term it modifies.

☐ Verb tenses are at least locally consistent.

☐ The words *due to* appear in the predicate and modify the subject.

☐ Correlative conjunctions such as *neither/nor* are properly used.

☐ Prepositions are carefully selected throughout.

☐ All numbers are expressed in standard fashion.

☐ Numerical values are paired with units where appropriate.

☐ Enumerations and lists are formatted properly.

☐ Smooth transitions are implemented throughout.

☐ Suitable words (technical and nontechnical) are employed throughout.

☐ All crucial terms are suitably defined.

The following should be *absent*.

☐ logical fallacies

☐ sentence fragments

☐ run-on sentences

☐ dangling verbals

☐ squinting modifiers or other forms of ambiguity

☐ vagueness

☐ misspelled words

☐ misused logical indicators

☐ first-person "I" or "me"

☐ gender-biased language

☐ colloquialisms, internet slang, "simplified spelling"

☐ unexpanded or unnecessary acronyms

☐ unnecessary non-English expressions

☐ rhetorical questions

☐   "different than"

☐   awkward constructions such as "of the … of the … of the …"

☐   unnecessary repetition, leftover scaffolding

☐   lazy, indefensible assertions

☐   cluttered or busy visuals

Eighteen Ways to Start a Sentence

The list below is adapted from *Grammar for Journalists* by E.L. Callihan (see p. 151 for the full reference). Although engineering is not journalism, writers in both professions must be direct while still varying sentence structure enough to keep the reader's interest. We mostly retain Callihan's headings but provide our own examples.

1. (**noun**) *Transistors* can amplify signals.

2. (**prepositional phrase**) *Upon simplification* we obtain $x^2 = y$.

3. (**present participial phrase**) *Using the quadratic formula*, we get $x = 1$ and $x = -2$.

4. (**past participial phrase**) *Encouraged by the success of this design*, the team continued to pursue the Euler approach.

5. (**infinitive phrase**) *To approach the problem more directly*, we adopted a computer-aided design methodology.

6. (**noun clause**) *Whether the problem is solvable* remains an open question.

7. (**subordinate clause of cause**) *Because $x_0 \geq 2$*, we have $f(x_0) = x_0^2 \geq 4$.

8. (**subordinate clause of concession**) *Although the machine failed*, it provided the designers with valuable information.

9. (**subordinate clause of condition**) *If $x_0 \geq 2$*, then $f(x_0) = x_0^2 \geq 4$.

10. (**subordinate clause of time**) *As its cooling unit began to fail*, so did the whole system.

11. (**gerund**) *Teaching* is difficult when the students lack sufficient motivation.

12. (**nominative absolute**) *Its main engine destroyed*, the plane started a rapid emergency descent.

13. (**verb**) *Solve* the problem in the usual manner.

14. (**adjective**) *Simple and rugged*, the machine served its purpose for 22 years.

15. **(adverb)** *Finally*, we take $x_0 = 2$.

16. **(pronoun)** *None* of these values satisfies the equation exactly.

17. **(expletive)** *There is* another, equally valid, solution to this problem.

18. **(conjunction)** *But* $x_0 > 2$, hence $f(x_0) = x_0^2 > 4$.

See the *Quick-and-Dirty Grammar Glossary* (p. 167) for some discussion of grammatical terminology.

Standard Abbreviations

abbreviation	meaning
etc.	and other things
viz.	namely
e.g.	for example
i.e.	that is
cf.	compare
n.b.	note well
et al.	and others
pp.	pages
ff.	and the pages following

Greek Alphabet

alpha	A	α	nu	N	ν
beta	B	β	xi	Ξ	ξ
gamma	Γ	γ	omicron	O	o
delta	Δ	δ	pi	Π	π
epsilon	E	ϵ	rho	P	ρ
zeta	Z	ζ	sigma	Σ	σ
eta	H	η	tau	T	τ
theta	Θ	θ	upsilon	Y	υ
iota	I	ι	phi	Φ	ϕ
kappa	K	κ	chi	X	χ
lambda	Λ	λ	psi	Ψ	ψ
mu	M	μ	omega	Ω	ω

SI (MSKA) Units and Abbreviations

Base Units

quantity	unit	abbreviation
length	meter	m
mass	kilogram	kg
time	second	s
electric current	ampere	A
thermodynamic temperature	kelvin	K
amount of substance	mole	mol
luminous intensity	candela	cd

Some Derived Units

quantity	unit	abbreviation
area	square meter	m^2
volume	cubic meter	m^3
speed, velocity	meter per second	m/s
acceleration	meter per second squared	m/s^2
electric current density	ampere per square meter	A/m^2
planar angle	radian	rad
solid angle	steradian	st
frequency	hertz	Hz
force	newton	N
pressure, stress	pascal	Pa
energy, work	joule	J
power	watt	W
electric charge	coulomb	C
electric potential	volt	V
electrical capacitance	farad	F
electrical resistance	ohm	Ω
electrical conductance	siemens	S
magnetic flux	weber	Wb
magnetic flux density	tesla	T
inductance	henry	H
Celsius temperature	degree Celsius	$^\circ$C

A more complete list of derived units appears in *The International System of Units (SI)*, NIST Special Publication 330, National Institute of Standards and Technology, 2008.

Unit Prefixes

prefix	abbreviation	numerical factor
exa	E	10^{18}
peta	P	10^{15}
tera	T	10^{12}
giga	G	10^{9}
mega	M	10^{6}
kilo	k	10^{3}
milli	m	10^{-3}
micro	μ	10^{-6}
nano	n	10^{-9}
pico	p	10^{-12}
femto	f	10^{-15}
atto	a	10^{-18}

Helpful Introductory Phrases

For example, ...

On the other hand, ...

As a special case, ...

For instance, ...

More importantly, ...

Indeed, ...

On the contrary, ...

Furthermore, ...

Despite this, ...

In conclusion, ...

In contrast, ...

Finally, ...

Nevertheless, ...

Without loss of generality, ...

Moreover, ...

However, ...

Strictly speaking, ...

Conversely, ...

Fluff Phrases

the reason is because

all things being equal

on the right track

as a matter of fact

easier said than done

better safe than sorry

it stands to reason that

last but not least

for the purpose of

needless to say

as we move forward

at this point in time

we might add that

it is noteworthy that

a comment is in order

as regards

it may be said that

it might be stated that

Examples of Redundant Expressions

~~absolutely~~ certain	~~end~~ result	~~absolutely~~ necessary
~~final~~ outcome	cancel ~~out~~	joined ~~together~~
~~clearly~~ obvious	reason ~~why~~	combine ~~into one~~
~~still~~ remains	~~completely~~ eliminated	~~totally~~ eliminated
each ~~and every~~	~~very~~ unique	earlier ~~in time~~
whether ~~or not~~	~~perfectly~~ straight	~~very~~ obvious
~~rather~~ shallow	~~quite~~ rapid	~~rather~~ hard
~~most~~ extraordinary	~~somewhat~~ important	~~always~~ necessary
~~particularly~~ striking	~~final~~ completion	combine ~~together~~
~~fine~~ line	~~highly~~ significant	~~wide~~ variety
~~serious~~ crisis	~~future~~ planning	~~valid~~ information
~~completely~~ surrounded	continue ~~on~~	rarely ~~ever~~
later ~~on~~	~~past~~ history	~~past~~ experience
~~whole~~ system	recur ~~again~~	revert ~~back~~
~~mutual~~ cooperation	in ~~actual~~ fact	if ~~and when~~
~~totally~~ destroyed	~~general~~ rule	~~particularly~~ relevant
~~wholly~~ new	~~hard~~ evidence	

Examples of Words Requiring Certain Prepositions

accede *to*	accordance *with*	adapted *to*
adjacent *to*	analogous *to*	applicable *to*
appropriate *to*	associated *with*	attach *to*
attribute *to*	apart *from*	belong *to*
characterized *by*	coincide *with*	compare *to* or *with*
conform *to*	connect *with*	consist *of* or *in*
comply *with*	consult *with*	contiguous *to*
contrary *to*	contrast *with*	decide *on*
depend *on* or *upon*	dependent *on* or *upon*	difficulty *in*
disagree *with*	distinct *from*	equivalent *to*
exception *to*	favorable *to*	impart *to*
impose *on* or *upon*	inconsistent *with*	independent *of*
indicative *of*	insist *on* or *upon*	made *of*
necessary *to* or *for*	noncompliance *with*	opposite *to*
peculiar *to*	perpendicular *to*	pertain *to*
prepare *for*	proportion *to*	recoil *from*
relate *to*	replete *with*	resort *to*

respond *to*
suitable *to* or *for*
terminate *with*
unequal *to*

substitute *for*
sufficient *for*
tied *to*
vary *in* or *with*

similar *to*
tend *to* or *toward*
unable *to*

British vs. American Spelling

British	American
analyse	analyze
catalogue	catalog
centre	center
colour	color
criticise	criticize
endeavour	endeavor
fibre	fiber
focussed	focused
labelled	labeled
litre	liter
metre	meter
organisation	organization
recognise	recognize
travelled	traveled

Signpost Headings

Aim	Algorithm	Alternate Route
Analogy	Analysis	Approach
Assertion	Assumption	Background
Basis	Caution	Claim
Clarification	Comment	Conclusion
Confirmation	Convention	Corollary
Criterion	Definition	Demonstration
Derivation	Description	Details
Detour	Example	Explanation
Fact	Goal	Guideline
Hypothesis	Idea	Illustration
Intent	Interpretation	Justification
Lemma	Limitations	Motivation
Method	Notation	Note

Objective	Observation	Pattern
Plausibility Argument	Premise	Principle
Problem	Procedure	Process
Proof	Proposition	Purpose
Question	Rationale	Reason
Recommendation	Remark	Restrictions
Result	Review	Rule
Shortcut	Solution	Starting Point
Steps	Strategy	Summary
Tactic	Technique	Test
Theorem	Trick	Typical Case
Validation	Verification	Warning

Words Often Confused

Refer to a dictionary to avoid confusion over words such as the following.

already, all ready	threw, through
their, there, they're	lead, led
forth, fourth	coarse, course
lightning, lightening	its, it's
deprecate, depreciate	device, devise
loose, lose	seams, seems
precede, proceed	decent, descent, dissent
sit, set	accept, except
advice, advise	to, too, two
we're, where, were	your, you're
accede, exceed	adherents, adherence
affluence, effluence	allowed, aloud
allusion, illusion	appressed, oppressed
assistance, assistants	band, banned
capital, capitol	censor, sensor
cash, cache	celery, salary
compliment, complement	deference, difference
emigrate, immigrate	eminent, imminent, immanent
enumerable, innumerable	forward, foreword
enquire, inquire	incredible, incredulous
hire, higher	hole, whole
ingenious, ingenuous	overdo, overdue
passed, past	plain, plane

cue, queue	rap, wrap
right, write	sight, site, cite
straight, strait	waist, waste
exceptional, exceptionable	distinct, distinctive
likely, liable	negligence, neglect
healthy, healthful	interpolate, extrapolate
farther, further	lay, lie
adapt, adopt	all together, altogether
credulity, credibility	deficient, defective
weather, whether	principal, principle
personal, personnel	moral, morale
peace, piece	stationary, stationery

Idiomatic Replacements

The term *idiom* has several possible senses, but in this case refers to a group of words whose net meaning is something other than the mere sum of the meanings of its constituent words. The following sorts of substitutions can eliminate idiomatic phrases and add directness to your writing.

replace		with	
	agree to	with	accept, approve
	back of	$\longrightarrow$	behind
	bring out		reveal, show
	bring about		accomplish, obtain, produce
	call for		demand, require
	carry out (or carry on)		perform, conduct
	come together		converge, meet
	deal with		treat, discuss
	decide on		select
	dispense with		omit
	do away with		discard
	end up		conclude
	fall off		decline, decrease
	fill out		complete
	fix up		repair, organize
	go into		investigate, examine
	go on with		continue
	keep out		exclude
	keep up		maintain
	look for		anticipate, expect

look upon	$\longrightarrow$	regard
make up		compose
make use of		utilize
out loud		aloud
plug in		substitute
put in		insert
put up with		endure
refer to		mention
show up		appear, arrive
size up		estimate, judge
speed up		accelerate, hasten
throw out		discard
try out		test
work out		devise, develop

(Based upon lists in *Technical Writing* by Rickard and *Handbook of Better Technical Writing* by Rowland. See p. 147 for the full references.)

Non-English Expressions

This category of expressions was discussed on p. 99.

a fortiori — by a stronger reason, all the more

a priori — from what is before

ad hoc — unplanned, done only when necessary

ad infinitum — endlessly

ad nauseam — to an excessive (literally, sickening) degree

ad interim — in the mean time

apropos — with reference/regard/respect to, concerning

bona fide — in good faith, true, not intended to deceive

circa — about, approximately

de facto — in practice, in effect, in reality, in fact

en masse — as a whole

en route — on the way, along the way

et alia (et al.) — and others

et cetera (etc.) — and other things

et sequentia (et. seq.) — and what follows

fait accompli — an accomplished fact, something already done

faux pas — mistake (false step)

ibidem (ibid.) — in the same work

in memoriam — in memory of

in situ — in position

in toto — in full

inter alia — among other things

ipso facto — by the fact itself

loco citato (loc. cit.) — in the place cited

magnum opus — chief work of an author

modus operandi — typical mode of operating

mutatis mutandis — the necessary changes having been made

opere citato (op. cit.) — in the work cited

per diem — daily

per se — as such

petitio principii — begging the question (fallacy)

pro rata — proportionately

quid pro quo — something for nothing

reductio ad absurdum — reduction to absurdity (a proof method)

sine qua non — something indispensable or necessary

sub verbo (s.v.) — under the word

versus (vs.) — against

via — by way of

Caution: Publishers vary on whether terms such as these should be italicized.

Quick-and-Dirty Grammar Glossary with Examples

Our intention here is to keep things simple. Please refer to a standard book on English grammar for more rigorous definitions.

Adjective

Adjectives modify nouns. In the sentence

> The sensor provides raw data.

the word *data* is a noun modified by the adjective *raw*.

Adverb

Adverbs modify verbs, adjectives, and other adverbs. In the sentence

> Subsystem *A* failed rapidly as the temperature exceeded the boiling point.

the word *rapidly* is an adverb modifying the verb *failed.*

Agreement in Number

English words can change form according to whether they refer to the *singular* (one thing) or the *plural* (more than one thing). Consider the sentences

> Subsystem *A* was functioning normally.
>
> Subsystems *A* and *B* were functioning normally.

The first sentence makes a statement about one thing (Subsystem *A*) via the singular verb form *was*. The second sentence makes a statement about two things (Subsystems *A* and *B*) via the plural verb form *were*. Now consider

> Subsystem *A* was functioning normally; it required no further investigation.
>
> Subsystems *B* and *C* were functioning normally; they required no further investigation.

The words *it* and *they* are pronouns. The singular form *it* must be used to refer back to the singular *Subsystem A*. The plural form *they* must be used to refer back to the plural *Subsystems B and C*. These are examples of *agreement in number* in the English sentence. Here's a catastrophic failure to implement agreement in number:

> The reference [4] gives a precise definition of the various quantities that appears in these equations.

In order to fix this sentence, this author must decide what he or she means to say. As readers, we might key on the plural nature of *various quantities* and interpret the sentence as

The reference [4] gives precise definitions of the various quantities that appear in these equations.

However, guessing at how to fix the sentence shouldn't be our job.

Ambiguity

Ambiguity is the possibility of more than one meaning. The sentence

Link A will overheat and fail only if switch S closes.

is ambiguous. It could mean either of the two different statements

[A will overheat] and [if A fails then S closes].
If [A overheats and fails] then S closes.

Articles

In English, the words *a*, *an*, and *the* are called the articles. These words act as modifiers. Some languages, such as Russian, do not have articles.

Clause

A clause is a group of words containing a verb and its subject and constituting part of a sentence. In the sentence

The current is too large, but the voltage is too small.

we have two clauses: (1) *The current is too large*, and (2) *the voltage is too small*. Each clause expresses a complete thought and has both a subject and a predicate. In the sentence

We have $y = ax^2$ when both b and c vanish.

the *main, principal,* or *independent clause* is *We have* $y = ax^2$. The rest of
the sentence, *when both b and c vanish,* is a *subordinate* or *dependent clause*.
A subordinate clause functions as a single part of speech (and cannot stand
on its own). In the sentence

> When the unit overheated, it failed.

the principal or main clause is *it failed*. Despite its simplicity, this clause
could stand on its own. The clause *When the unit overheated* is subordinate;
it cannot stand on its own.

Conjunction

Conjunctions are grammatical connectors. *Coordinating conjunctions* (such as
and, but, either/or, and *neither/nor*) join coordinate clauses or independent
clauses. *Subordinating conjunctions* (such as *that, after, because, though,* and
if) join subordinate clauses to principal clauses. In the sentence

> Unit *A* failed before the process was complete.

the conjunction *before* joins the subordinate clause *before the process was
complete* to the principal clause *Unit A failed*.

Dangling Modifier

A adjective-type modifier dangles when the word it's supposed to modify is
missing. In the sentence

> By adjusting the settings, the unit can be configured in many
> different ways.

the agent (entity who will adjust the settings) is missing. A correct version is

> By adjusting the settings, we can configure the unit in many dif-
> ferent ways.

Gerund

Gerunds are words constructed from verbs but used as nouns. Consider the sentences

> Measuring is the best way to learn about voltage.
>
> Designing for the optimum is not always feasible.

In the first one, the word *measuring* is a gerund. In the second one, *Designing for the optimum* is a *gerund phrase* Gerunds always have the *-ing* ending. Compare *participle* on p. 173.

Idiom

See p. 165.

Imperative Mode

The imperative mode expresses a command, desire, or permission. In the sentence

> Let us investigate this further.

the verb *investigate* is in the imperative mode. Compare *indicative mode*, *subjunctive mode*.

Indicative Mode

The indicative mode expresses a declaration. It appears in such sentences as

> The channel can introduce distortion and noise.
>
> The phase of the signal is not predictable.

Compare *imperative mode*, *subjunctive mode*.

Infinitive

An infinitive is a verb form starting with the word *to* (e.g., *to inspect*). Such a form can be used as a noun, an adjective, or an adverb. Consider the sentences

> To err is human.
>
> This is a phenomenon to observe.
>
> A team of engineers was sent to inspect the site.

In the first one, the infinitive *To err* is the subject of the sentence. In the second one, *to observe* functions as an adjective modifying *phenomenon*. In the third one, *to inspect* functions as an adverb modifying the verb *was sent*. A *split infinitive* is formed when an adverb is inserted between the the word *to* and the rest of an infinitive:

> It is impossible to rapidly prototype such a system.

Intransitive Verb

An intransitive verb does not require an object to complete its meaning. The sentence

> Such an antenna radiates nonuniformly.

contains the intransitive verb *radiates*. Contrast with *transitive verb* on p. 178.

Metaphor

A metaphor is an unacknowledged comparison. The sentence

> The invention of calculus was the final dagger through the heart of trial and error as a design methodology.

uses metaphor to communicate an idea. Calculus was not *really* a dagger (in the weapon-related sense), and trial and error does not *really* have a heart (in the biological sense). A comparison was made but not explicitly acknowledged. Contrast with *simile*.

Mode (or Mood)

See *indicative mode, subjunctive mode, imperative mode.*

Noun

Nouns name things. A *common noun* names any element of a class. A *proper noun* distinguishes something from the other elements of the class to which it belongs. Consider the sentences

> Each chapter ends with a number of exercises.
>
> A discussion of this phenomenon appears in Chapter 2.

Here *chapter* is a common noun (note that it is not capitalized). *Chapter 2* is a proper noun (note that it is capitalized); it denotes a particular chapter in the book or report.

Participle

Participles are words constructed from verbs and used as adjectives. Consider the sentences

> A working subsystem is desired.
>
> Substituting into equation (6), we find that $x = 4$.

In the first one, the word *working* is a participle; it is formed from the verb *work* and used as an adjective to modify the noun *subsystem*. In the second one, the participle *substituting* begins a *participial phrase* which modifies the pronoun *we*. These *-ing* participles are called *present participles*. The following sentences contain examples of *past participles*:

> Let us consider the proposed expression.
>
> The solution found in Chapter 1 will play an essential role here.
>
> Weakened by years of neglect, the cable snapped during a storm.

Parts of Speech

English words are classified into *parts of speech* according to the functions they perform in sentences. The eight traditional parts of speech are tabulated below.

part of speech	function
noun	names a person, place, or thing
pronoun	takes the place of a noun
adjective	modifies a noun or a pronoun
verb	expresses action or state of being
adverb	modifies a verb, an adjective, or another adverb
preposition	shows the relation of a noun or a pronoun to another word in the sentence
conjunction	joins words or groups of words
interjection	expresses emotion

Phrase

A phrase is a group of words functioning as a grammatical unit (compare with *clause*). For example, a *noun phrase* is a group of words functioning as a noun. In the sentence

> The lumped element approximation ignores the essential wave nature of electromagnetic phenomena and is therefore useful only at sufficiently low frequencies.

we can identify the noun phrases *The lumped element approximation*, *the essential wave nature of electromagnetic phenomena*, and *sufficiently low frequencies*. But there are also prepositional phrases, verb phrases, etc., in English.

Preposition

A preposition is used to introduce a noun or pronoun and establish its relationship to something else. In the sentence

> Next, we solved for the field values in the excluded regions.

the word *in* is a preposition. The phrase *in the excluded regions*, beginning with a preposition and ending in a noun (the *object* of the preposition) is a *prepositional phrase*. Other examples of prepositions are *after*, *before*, *from*, *under*, *toward*, and *with*.

Pronoun

Pronouns take the place of nouns (i.e., they represent objects or people without naming them). The noun to which a pronoun refers is called the *antecedent* or *referent* of the pronoun. In the sentence

> Although the ideal low-pass filter is not realizable, it is often used as a comparison for describing the properties of real filters.

the word *it* is a pronoun taking the place of the noun *the ideal low-pass filter*. In the sentence

> Although f is complex, it is independent of time t.

the antecedent of the pronoun *it* is the mathematical quantity f.

Sentence

A sentence expresses (or implies) a complete thought. A *declarative sentence* asserts a fact or affirms a proposition. The sentence

> The weight W of an object is given by $W = mg$.

is declarative. The *subject* (the thing written about) is *The weight W of an object*. The rest is the *predicate* (that which is asserted about the subject). In the sentence

> This book provides a thorough introduction to antenna theory.

the phrase *provides a thorough introduction to antenna theory* is the predicate. It asserts something about the subject of the sentence, which is *This book*.

Sentence Fragment

A sentence fragment is a portion of a sentence punctuated as though it were a sentence. Consider the fragments

Which shows that $x = 2$.

After the analysis was completed.

Substituting into equation (1).

Simplified by factoring.

The first two are subordinate clauses. The last two are participial phrases. None is a sentence.

Simile

A simile is an acknowledged comparison, introduced in English by *like*, *as*, or *so*. The sentence

Electromagnetic waves are like waves on water.

uses simile to communicate an idea. Contrast with *metaphor* on p. 172.

Squinting Modifier

An adverb squints when sitting between two verbs. In the sentence

The system we had designed rapidly failed on first use.

the adverb *rapidly* squints because it could be applied to either of the verbs *designed* or *failed*. Did they design the system rapidly, or did it rapidly fail? See *ambiguity*.

Subjunctive Mode

The subjunctive can be used to express a condition, motive, or supposition. It is introduced by such words as *if*, *though*, *unless*, *except*, *whether*, *that*, and *provided*. We suggested you use the subjunctive for conditions contrary to fact; it is also suitable for expressing conditions that are merely improbable. Compare *indicative mode*, *imperative mode*.

Tense

Tense is a verb modification used to indicate time of action. English has six tenses. The *present tense* communicates something now happening or existing, something typical or habitual, or something true at all times:

> We wish to determine x.
>
> Our engineers design microwave circuits.
>
> Several other arguments reinforce this view.

For historical reasons, it is sometimes used for the future:

> Our company declares bankruptcy next month.

The *past tense* communicates something happening, existing, or recurring in the past:

> Newton developed an equation to describe the effects of gravity.
>
> Nineteenth-century physicists were seldom rewarded for their work.

The *future tense* communicates something expected to happen. It is signaled by the words *shall* or *will*:

> Subsystem A will fail before the end of the month.

The three *perfect* tenses describe an action as completed at some point in time. The *present perfect tense* describes something begun in the past and completed by the present time. It is signaled by the words *has* or *have*:

> We have demonstrated that $x = a/2$.

The *past perfect tense* describes something that occurred in the past before something else occurred. It is signaled by the word *had*.

> Smith had completed his dissertation before the development of general relativity.

The *future perfect tense* describes something expected to occur in the future before something else occurs. It is signaled by the words *will have*:

> Jones will have completed his report before next summer.

Attention to tense is required to maintain the actions in a sentence in proper time relation to one another. The sentence

> Our research effort provided results and came in under budget, but it proceeds too slowly.

exhibits a *shift* in tense from past to present. Unnecessary shifts should be avoided.

Transitive Verb

A transitive verb is one whose action passes over to an object. The sentence

> Section 2.2 describes the performance of *L*-type machines.

contains the transitive verb *describes*. Contrast with *intransitive verb*.

Verb

Verbs say things about things; they assert action, being, or condition. In the sentence

> We define x as the distance between points A and B.

the word *define* is a verb. Its *subject* (the thing acting) is *We*. Its *object* (the thing acted upon) is the mathematical quantity x. In the sentence

These systems gather information.

the verb is *gather* and its object is *information*. A verb can consist of more than one word. In the sentence

The unit *was replaced* at that time.

the italicized words form a *verb phrase*. See *mode, tense, voice, intransitive verb, transitive verb, verbal.*

Verbal

See *gerund, infinitive,* and *participle.*

Voice

Voice is a verb modification indicating whether the subject performs the action or receives the action. In the *active voice*, the subject performs the action. In the *passive voice*, the subject receives the action. In the sentence

Transistor A requires a power source to operate.

the subject (*Transistor A*) performs the action (it *requires* something). This is the active voice. In the sentence

The electron was set in motion by an electric field.

the subject (*electron*) receives the action (of being set in motion). This is the passive voice. The active voice can lend directness to a statement and is often recommended by books on writing.

Index

emphasis, viii, 32
empirical–inductive viewpoint, 60
empty set, 145
en masse, 166
en route, 166
engineering design process, 3, 8
    example, 3
    flowchart, 3
    steps, 3
engineering, definition of, 21, 142
enumerations, 78
equals sign, 116, 119
equation, 105
    breaking long, 116
    displayed, 117
    inline, 117
equivocation, fallacy of, 47, 156
et al. (abbreviation), 70, 159, 166
et alia, 166
et cetera, 166
et sequentia, 166
et. seq. (abbreviation), 166
etc. (abbreviation), 70, 159, 166
ethics, 19
evidence, 54
exa (prefix), 161
exclamation point, 65
existence of solution, 135, 141
existence/uniqueness quantifier, 136
existential quantifier, 120, 128
experimental methodology, 34
exposition, 24
expression, 105, 112

fait accompli, 166
fallacies, 36, 42, 55, 155
fallacy of accident, 33, 43, 155
fallacy of amphiboly, 47, 156
fallacy of composition, 45, 156
fallacy of division, 45, 156
fallacy of equivocation, 47, 156
fallacy of opposition, 48, 156
false dichotomy, 46, 156
farther vs. further, 81
faux pas, 166

feedback, getting, 7, 9, 104
femto (prefix, 161
fewer vs. less, 81
ff. (abbreviation), 70, 159
Fibonacci numbers, 145
figure callouts, 104
figures, 101
first person, 96
fluff phrases, 68, 79, 161
foreign phrases, 99
form and function, 12
formal document, viii
formal outline, 28
formality, 17, 21, 105
fractions, typesetting, 116
frogs, 27
function, 112
further vs. farther, 81
future perfect tense, 178
future tense, 177

garbage in, garbage out, 20
gender-neutral language, 96
general truth, 74
generality, 137
generalization, hasty, 44
genus, 124
gerund, 74, 171
gerund phrase, 171
giga (prefix), 161
grammatical prominence, 63
graph, 113
Greek alphabet, 101
grouping symbols, 120

hasty generalization, 44, 155
heuristic reasoning, 151
hybrid design approach, 6
hyphen, 66, 67

I (first person), 95
i.e. (abbreviation), 70, 159
ibid. (abbreviation), 166
ibidem, 166
identity, 113
idiom, 165, 171

RECEIVED

DEC 1 8 2014